AF714561

COLLECTIVE CLIMATE ACTION AND NETWORKED CLIMATE GOVERNANCE

Edited by

Dr. Pralhad Jaybhaye

Delve
PUBLISHING

www.delvepublishing.com

Collective Climate Action and Networked Climate Governance

Dr. Pralhad Jaybhaye

Delve Publishing
224 Shoreacres Road
Burlington, ON L7L 2H2
Canada
www.delvepublishing.com
Email: orders@arclereducation.com

© 2024 Delve Publishing

ISBN: 978-1-77469-947-8 (Hardcover)

This book contains information obtained from highly regarded resources. Reprinted material sources are indicated and copyright remains with the original owners. Copyright for images and other graphics remains with the original owners as indicated. A Wide variety of references are listed. Reasonable efforts have been made to publish reliable data. Authors or Editors or Publishers are not responsible for the accuracy of the information in the published chapters or consequences of their use. The publisher assumes no responsibility for any damage or grievance to the persons or property arising out of the use of any materials, instructions, methods or thoughts in the book. The authors or editors and the publisher have attempted to trace the copyright holders of all material reproduced in this publication and apologize to copyright holders if permission has not been obtained. If any copyright holder has not been acknowledged, please write to us so we may rectify.

Notice: Registered trademark of products or corporate names are used only for explanation and identification without intent of infringement.

Delve Publishing publishes wide variety of books and eBooks. For more information about Delve Publishing and its products, visit our website at www.delvepublishing.com.

ABOUT THE EDITOR

Dr. Pralhad Jaybhaye (1972) is presently serving as Associate Professor, Department of Agriculture Meteorology, Vasantrao Naik Marathwada Krushi Vidyapeeth, Parhani, (Maharashtra). He did his B. Sc.(Agri.) in 1995 and M. Sc. (Agriculture Meteorology) 1998 from Mahatma Phule Krushi Vidyapeeth, Rahuri (College of Agriculture, Pune) and Ph. D. (Agril. Meteorology & Physics) in 2020 from BCKV, Mohanpur, West Bengal. He is recognized a Post Graduate Teacher and Research Guide for M. Sc. And Ph. D. (Agriculture Meteorology). He started his career as a Technical Officer (Agrometeorology), NARP, Igatpuri (MPKV, Rahuri) from 2000 to 2008. His field of specialization is Agrometeorology and his area of research interest are Micrometeorology, Agricultural Water Management, Climate Change and Extension Agrometeorology. In addition to this, he has performed additional duties as a Nodal officer, Commercial Fruit Nursery, PDKV, Akola (2005-2008); Principal Nodal officer, GKMS & FASAL project (2008-2016) and Principal, Agriculture Technology School, Jalna (VNMKV, Parbhani) (2020-2023). He had conducted National workshop, farmer training programs, workshops and many farmer's rallies/awareness programmes. During his total 25 year's tenure of services, under his guidance 27 students completed M.Sc. (Agrometeorology) and he has co-guided for more than 25 students to the same degree course. He has published one Text Book (Handbook of Agrometeorology); 2 Subject Books (1-Hydrometeorology, 2-Weather and Climate Monitoring and Prediction Systems in Context of Agriculture) and other 6 books related to Agrometeorology; 25 research papers & 05 short communications in international and national reputed journals. He has participated and presented research papers more than 25 International, National, State Seminar, Workshops and Symposiums. Besides these he had done extension Agricultural Meteorology activities like TV talk (17) & Radio talks (200), guest lectures in scientist training & farmer training programs had delivered. Dr. Pralhad Jaybhaye has also written more than 450 popular articles in reputed regional language periodicals and daily newspapers. He had donated Gold Medal to M. Sc. (Agriculture Meteorology), VNMKV, Parbhani in the memory of his father (Late Freedom Fighter Mr. Rambhaji Temaji Jaybhaye) for enrichment capacity building as well as raise academic interest in the Agriculture Meteorology area.

TABLE OF CONTENTS

LIST OF FIGURES

LIST OF ABBREVIATIONS

CBDR-RC	Common but Differentiated Responsibilities and Respective Capabilities
CDM	Clean Development Mechanism
CH_4	Methane
CO_2	Carbon Dioxide
COP	Conference of the Parties
ETS	Emission Trading Schemes
FFF	Fridays for Future
GCCA	Global Climate Change Alliance
GCF	Green Climate Fund
GEF	Global Environment Facility
GHG	Greenhouse Gas
ICAO	International Civil Aviation Organization's
IEA	International Energy Agency
IET	International Emissions Trading
IKI	Germany's International Climate Initiative
IPCC	Intergovernmental Panel on Climate Change
JI	Joint Implementation
NAMAs	Nationally Appropriate Mitigation Actions
NAPs	National Adaptation Plans
NDCs	Nationally Determined Contributions
PPPs	Public-Private Partnerships
REEEP	Renewable Energy and Energy Efficiency Partnership
SBI	Subsidiary Body for Implementation

UNDP	United Nations Development Program
UNFCCC	United Nations Framework Convention on Climate Change
WBCSD	World Business Council for Sustainable Development
WRI	World Resources Institute

PREFACE

The urgency of combating climate change has never been more apparent. With unprecedented weather events, rising sea levels, and rapidly melting ice caps, the fight to protect our planet is an immediate and shared responsibility. The book, "Collective Climate Action and Networked Climate Governance," is an essential guide that explores this fight, presenting a comprehensive overview of current climate governance and discussing the path forward. With ever-growing acknowledgment of the interconnectedness of our world, the book is a timely exploration of how collective action and networked governance can become key drivers of climate action.

This book is structured around a set of broad themes that permeate the landscape of climate governance. These themes include understanding the historical development of climate policy, the role of non-state actors in addressing climate change, the effectiveness of international climate agreements, and the potential of technology and innovation to shape our climate future. Each theme is intricately linked to the underlying question of how collective action and networked governance can be harnessed to accelerate the shift toward sustainable and resilient societies.

In exploring these themes, the book delves into the historical aspects of climate governance, uncovering the origins of the climate change debate and the evolution of policy responses to environmental degradation. This historical account provides readers with a foundational knowledge of climate governance, as well as highlighting the progression and shortcomings of past climate policies. Thus, it emphasizes the necessity of transcending the limitations of traditional policy-making approaches, making space for innovative governance structures and cross-sector collaborations.

In an increasingly globalized world, the book also provides a detailed analysis of the contemporary landscape of climate governance. It identifies the complex interplay of various actors, including states, transnational organizations, cities, and businesses, in shaping climate policy. By shedding light on these varied stakeholders, the book demonstrates the immense potential of networked climate governance to drive effective and transformative collective action. It highlights the power of synergistic partnerships and the importance of integrating the different dimensions of climate action – mitigation, adaptation, and the management of loss and damage.

Looking to the future, the book outlines emerging trends and innovative approaches to climate governance. It discusses how digital technologies, such as blockchain and artificial intelligence, may revolutionize our capacity to monitor, verify, and enforce climate commitments. Furthermore, it examines the prospects of adopting circular economy models and harnessing decentralized renewable energy systems that promote a just and inclusive transition to a low-carbon world. By showcasing these forward-thinking possibilities, the book encourages readers to assume an active role in shaping the future trajectory of climate governance.

"Collective Climate Action and Networked Climate Governance" has been designed to appeal to a wide range of audiences, making it an invaluable resource for undergraduate students, practitioners, policymakers, and the general public. For students, the book provides a comprehensive and accessible introduction to the field of climate governance, serving as an excellent foundation for further academic pursuit. For practitioners and policymakers, it offers insights into best practices, innovative approaches, and the potential of collaborative efforts, equipping them with the knowledge and tools to drive transformative change. Lastly, for the general public, the book serves as an engaging and informative read that raises awareness about the critical issues surrounding climate change and inspires individual action.

In conclusion, "Collective Climate Action and Networked Climate Governance" is a timely exploration of the complex landscape of climate governance. It highlights the potential of collaborative efforts, both on national and international scales, to tackle the pressing challenge of climate change and build more resilient societies. It is with the sincerest hope that readers enjoy this book and join the global movement in promoting collective climate action and innovative, networked governance.

CHAPTER 1

International Agreements on Climate Change

CONTENTS

As the existential threat posed by climate change to our planet becomes increasingly imminent, it has become impossible for humanity to sidestep the urgent call for substantive collective action. In an ever-globalizing world, addressing the issue of climate change necessitates the efficient coordination of both national and international strategies. Consequently, this chapter titled "International Agreements on Climate Change" seeks to analyze the role and significance of collective climate action and its embodiment within networked climate governance.

The recent upsurge in climate change activism has contributed significantly to the growing awareness of the imminent challenges that confront humanity. From shrinking ice caps to prolonged droughts and life-threatening floods, the need for immediate and decisive action has never been more discernible. As a direct result, international agreements on climate change—triggered by shared concerns for sustainable development and environmental conservation—have emerged as indispensable mechanisms for the coordination of global climate policies. These agreements stand as testament to the widespread recognition that climate change is not a localized problem but one that requires cooperation, solidarity, and synergy among nations.

Collective climate action refers to the coordinated efforts undertaken by multiple stakeholders, including governments, the private sector, and civil society organizations, to address and mitigate the adverse effects of climate change. This powerful approach helps share the burden of responsibility and resources needed to tackle the immense challenge, thereby broadening the scale of achievable environmental impact. Moreover, by promoting a collaborative network that transcends political, geographical, and ideological boundaries, collective climate action underscores the recognition that climate change presents an immense shared threat that can only be overcome through persistent and well-synchronized international collaboration.

Given the multidimensional nature of the climate change problem, traditional hierarchical governance structures may prove insufficient to mobilize and orchestrate the intricate and diverse array of actions required for a comprehensive response. Consequently, networked climate governance has emerged as an avant-garde approach to policy coordination and implementation in the context of international climate change agreements. Networked climate governance involves the formulation, negotiation, and execution of climate policies through the active interaction and collaboration between governments, international organizations, private sector entities, and civil society organizations. Essentially, networked climate governance captures the role of institutions that facilitate cooperation among multiple players by providing channels for information sharing, structuring incentives for coordination, and creating a sense of trust among participants.

In the present chapter, we will explore key international agreements on climate change that have fostered collective climate action and networked climate governance. Major milestones, such as the United Nations Framework Convention on Climate Change (UNFCCC), the Kyoto Protocol, and the Paris Agreement, will be critically examined and analyzed. Furthermore, specific issues and challenges that have arisen within these agreements and their implications for networked climate governance shall be thoroughly assessed. Finally, we will contemplate the future outlook of international collaboration and coordination within the climate change arena, seeking to elucidate potential future challenges and opportunities for bolstering the effectiveness of networked climate governance.

In unraveling the intricacies of international agreements on climate change, this chapter aims to convey a nuanced understanding of the significance of collective climate action and networked climate governance as indispensable components of an effective global climate strategy. As the clock ticks on humanity's quest for planetary salvation, the pivotal role of cooperation among nations in combating climate change becomes ever clearer—our collective destiny lies in our capacity to surmount the boundaries that divide us and embrace the unified strength that collaboration endows.

1.1. UNITED NATIONS FRAMEWORK CONVENTION ON CLIMATE CHANGE (UNFCCC)

The UNFCCC is a global environmental treaty initiated within the United Nations that provides a multilateral platform for countries to cooperate and collectively address the global issue of climate change (Teodoro et al., 2021; Therville et al., 2019; Tosun & Schoenefeld, 2017; Tuda & Machumu, 2019; van Bers et al., 2019). Signed in 1992 during the Earth Summit in Rio de Janeiro and coming into force in 1994, its main objectives revolve around stabilizing greenhouse gas (GHG) concentrations in the atmosphere, avoiding dangerous anthropogenic interference with the climate system, and promoting sustainable development within the context of reducing global emissions. The UNFCCC is an international agreement signed by 195 countries, which works to stabilize GHG concentrations in the atmosphere.

The purpose of this treaty is to prevent further global warming and reduce its adverse effects. To achieve this goal, UNFCCC calls for countries to adopt a range of measures that promote sustainable development and low-carbon growth.

Recent empirical evidence has shown that atmospheric GHG levels are continuing to rise despite efforts from the UNFCCC. According to the World Meteorological Organization's 2019 GHG Bulletin, carbon dioxide concentration has increased by 2% compared with 2018 values—reaching 407 parts per million in 2019—and continues rising at record speed due to human activities such as burning fossil fuels and deforestation. Methane concentrations have also risen sharply over recent years including from sources like agriculture and oil & gas production. Other potent gasses such as nitrous oxide have seen increases since pre-industrial times too; consequently contributing significantly towards climate change impacts worldwide. In order to meet their commitments under UNFCCC, many governments around the world are introducing policies aimed at reducing emissions while spurring economic growth through renewable energy investments or emission trading schemes (ETS). For example, the European Union introduced a new ETS scheme in 2021 allowing companies who produce high amounts of CO_2 emissions within certain sectors such as aviation or power generation trade among themselves so they can stay below their allocated quotas without having any additional costs imposed on them directly – creating financial incentives for businesses cut down on their emissions output instead of being punished financially if they do not comply with regulations regarding environmental protection standards set out by EU legislation. This move was made possible by collaboration between member states working together under the umbrella framework provided by UNFCCC guidelines – ultimately aiming stabilizing GHG concentration levels globally before it's too late for humanity address one of its biggest challenges yet (Figure 1.1).

The UNFCCC is a globally accepted agreement to reduce GHG emissions and avoid dangerous anthropogenic interference with the climate system (Rozance et al., 2019; Rudnick et al., 2019; Rutland & Aylett, 2008). The UNFCCC was created in 1992 as part of the Rio Earth Summit, which also gave birth to other prominent global environmental agreements such as Agenda 21 and the Kyoto Protocol. Its purpose is to protect the environment by stabilizing atmospheric concentrations of greenhouse gasses at levels that will prevent human-induced climate change from causing serious harm or destruction.

Figure 1.1. Summary of United Nations framework convention on climate change.

Source: Bashir Sir Political Science via YouTube.

Since then, international efforts have been made under this framework in order to address climate change, mitigate its impacts, and enhance national capacity for adaptation through various measures like strengthening scientific research capabilities around environmental issues; providing financial support for developing countries; promoting renewable energy usage over fossil fuels; investing in green infrastructure projects across nations; introducing carbon pricing mechanisms; increasing public awareness about ecological concerns among citizens etc.

In recent years there has been an increased push towards achieving net zero emissions within two decades given how pressing this issue has become due to accelerating rates of warming temperatures, melting glaciers and polar ice caps, rising sea levels along coastal areas threatening habitats for wildlife species, extreme weather events becoming more frequent resulting into humanitarian disasters etc., In fact according to data released by International Energy Agency's (IEA) 'Global Carbon Budget 2020' report total CO_2 emitted rose sharply between 2018–2019 after consistent dip during previous 3 years leading up 2019 indicating need for greater urgency while tackling this crisis especially when considering 2030 target set out under Paris Agreement aiming achieve 45% reduction compared 2010 emission level. Therefore, it can be seen as clear examples why avoiding dangerous anthropogenic interference with the climate system should remain a priority agenda item amongst signatories of UNFCCC even if progress being achieved so far falls short of expectations set against gravity challenges posed our times.

The UNFCCC was adopted in 1992 with the aim of stabilizing GHG concentrations in the atmosphere at levels that would prevent dangerous human interference with Earth's climate system (Goodwin et al., 2022; Harmon et al., 2023; Sanders et al., 2017; Yates, 2012; Zeren et al., 2022). To meet this goal, nations have committed to reducing their net GHG emissions by 2020 through Nationally Determined Contributions (NDCs). Studies show that the world has made progress towards meeting these goals since its adoption – reducing total annual GHG emission growth rates from around 3% annually between 2000–2010 to 1.7% between 2010–2018 – but further efforts are needed if we are going to stay within our 2°C temperature limit target set out in the Paris Agreement. According to official statistics, over half of all countries responsible for 85% of global emissions have already taken steps forward such as introducing carbon taxes or implementing renewable energy targets; however, much to be done before reaching sustainability objectives outlined under the convention.

Figure 1.2. Global governance structures for climate change management.

Source: Global Challenges foundation via Twitter.

For example, contemporary initiatives like 'green bonds' allow private investors direct access into green projects which helps finance new clean technologies and sustainably managed forests – two practices integral for promoting sustainable development and lowering global emissions according to scientists estimates. There has also been an increase in public awareness campaigns regarding how individuals can help contribute: In China alone nearly 500 million people took part in planting trees last year while many other countries around the globe turned off lights during specific hours as part of Earth Hour events designed to raise attention about environmental protection issues. Governments too should continue working together globally on policies that encourage low-carbon investment infrastructure and discourage fossil fuel subsidies– both practices essential for achieving long term success when it comes decreasing overall CO_2 output worldwide.

UNFCCC has been the global governance structure for climate change management since its adoption in 1992 (Figure 1.2). The UNFCCC is an international treaty that seeks to stabilize GHG concentrations and reduce emissions as a way of preventing further global warming. It sets out general principles, actions and objectives that all countries should follow in order to limit the effects of climate change (Fitch-Roy et al., 2020; Fried et al., 2022; Gregg et al., 2023; Shrestha et al., 2023; Therville et al., 2019). Recent years have seen more ambitious action taken by the UNFCCC with the Paris Agreement being signed in 2015 committing nations to collectively reducing their emissions by 2020. This was followed up with other initiatives such as NDCs which requires each country to submit plans outlining how they will meet their commitments, while also providing support for vulnerable states through access to technology transfers and capacity building opportunities. Moreover, scientific advances over recent decades have enabled better understanding into earth science systems including carbon cycles helping us understand our impact on them even further than before so we can take appropriate measures needed for mitigation or adaptation when necessary. For instance, using satellite imagery scientists are able to track deforestation rates across different regions allowing governments from around world who are party too parties of the convention work together towards a common goal of reducing emissions worldwide making sure no one gets left behind during this process either financially or technologically speaking due make it easier overall implementation success rate increase significantly compared previous efforts done prior entering agreement itself back then 2012–2013 period time frame where there were only few localized attempts combat issue rather than unified effort like now days currently ongoing right present moment 2021 onwards year current date today day month etc.

The signing of the UNFCCC represents a significant milestone in the recognition of anthropogenic climate change as a collective problem requiring holistic global action (Lee et al., 2014; Lee & Koski, 2014; Leventon et al., 2017). From as early as 1979, scientists have warned about the negative implications of rising global temperatures on socio-economic stability and ecological systems. Meanwhile, the collective efforts of governments, civil society, and the private sector under the UNFCCC aim to facilitate global, regional, and national initiatives to mitigate and adapt to climate change. Contemporary examples of collective climate action efforts under the UNFCCC umbrella include the establishment of the Green Climate Fund (GCF) and the Global Environment Facility (GEF). These finance mechanisms help developing countries access financial resources needed for climate change mitigation and adaptation projects. Additionally, the UNFCCC's many working groups and facilitative platforms, such as the Subsidiary Body for Implementation (SBI), offer technical support for the integration of climate policies into national development plans. Developed countries also play a vital role in collective climate action under the UNFCCC by providing financial, technological, and capacity-building support. Their commitments lead to fairer burden-sharing and the unlocking of resources for climate-resilient infrastructure and mitigation strategies in developing nations. For instance, the Paris Agreement under the UNFCCC mandates developed countries to mobilize $100 billion per year by 2020 for climate finance in developing nations. The financial flows assist in renewable energy projects, forest conservation initiatives, and early warning systems for extreme weather events.

The UNFCCC's framework encourages interdependent, transnational interactions in pursuit of climate objectives (Archer, 2016; Homobono et al., 2022; Ostrom, 2014; Panditharatne, 2016; Sara et al., 2016; Voogd et al., 2022). Such governance systems represent a growing network of non-state actors and sub-national governments that contribute to the global climate change agenda through decentralized, dynamic, and innovative approaches. City networks such as C40 Cities and the Global Covenant of Mayors exemplify contemporary examples of successful networked climate governance actors. These collaborations enable cities to share best practices, provide technical assistance, and conduct research on sustainable urban development initiatives. Through knowledge exchange, benchmarking, and collaboration, city networks empower their members to adopt low-carbon urban futures and maximize the collective climate change mitigation impact. Sectoral collaborations like the Renewable Energy and Energy Efficiency Partnership (REEEP) and International Civil Aviation Organization's (ICAO) CORSIA initiative offer cooperative roadmaps for the energy and aviation sectors to transition to sustainable industry practices. By creating partnerships between private enterprises, governments, and research institutions, these sectoral initiatives catalyze action across industries and facilitate unified responses to climate change challenges. Empirical evidence underscores the remarkable impact of networked climate governance initiatives in line with the UNFCCC's objectives. For instance, the IEA reported in 2020 that renewable energy capacity increased by 45% globally over the previous five years, driven by policy frameworks and regional initiatives. The growth indicates that sectoral collaboration and non-state actors' contributions have paced climate action progressively. Official statistics further demonstrate the importance of networked climate governance in the UNFCCC framework. The New Climate Institute, a leading climate policy research group, estimates that non-state and subnational initiatives under the Global Climate Action Agenda can reduce global GHG.

In conclusion, the UNFCCC has played a crucial role in shaping a global agenda that underscores and facilitates collective and networked approaches towards climate action and governance. Through its various mechanisms and platforms, the UNFCCC has promoted the equitable distribution of resources, skills, and knowledge-sharing across diverse stakeholders, encompassing governments, civil society, the private sector, and subnational entities. Moreover, the growing number of networked climate initiatives indicates a marked shift towards decentralized, dynamic, and innovative governance systems that collectively address the complex challenges of climate change. Empirical evidence and official statistics support the notion that these efforts have had a meaningful impact on global climate action. Building upon the foundation set by the UNFCCC, the international community further sought to establish binding commitments to reduce GHG emissions and promote sustainable development through the Kyoto Protocol. The protocol exemplifies the next phase in the global response to climate change, refining the mechanisms by which countries collaborate while also developing a legally binding framework to facilitate collective climate action. As a consequence of the UNFCCC's foundational groundwork, the progression towards robust, extensive, and coordinated global climate governance was evidenced in the breakthrough commitments of the Kyoto Protocol.

1.2. THE ROLE OF THE KYOTO PROTOCOL

The Kyoto Protocol, as a landmark international agreement on climate change, carries a legacy of cooperative global climate governance and paves the way to a new paradigm of multilateral environmental engagement (Davies et al., 2018; Rashchupkina, 2015; Shrestha et al., 2023; Vasconcelos et al., 2015). Born out of mounting global pressure concerning anthropogenic climate change, the Kyoto Protocol sought to build upon the progress made under the UNFCCC, as countries recognized the dire need for enforceable, binding commitments to reduce GHG emissions. The success of the Kyoto Protocol, however, has not been without controversy, as the treaty navigated the complexities of climate politics, diverse socio-economic interests, and uneven responsibility-sharing among developed and developing nations. The ensuing discussion on the Kyoto Protocol in relation to international agreements on climate change will delve into the treaty's origins, objectives, mechanisms, challenges, and implications for global climate governance. This comprehensive perspective aims to unveil the multifaceted nature of the Kyoto Protocol and its influence on shaping the foundations of contemporary climate policy and international cooperation. Born in the verdant heart of its eponymous Japanese city in 1997, the Kyoto Protocol marked a turning point as the first agreement under the UNFCCC to enforce quantified GHG emission reduction commitments, known as Annex B emissions targets, for industrialized countries (Annex I Parties). As a response to growing environmental concerns and a testament to the international community's willingness to address climate change, the treaty entered into force in 2005. By establishing legally binding GHG reduction targets, the Kyoto Protocol intensified global engagement and accountability, in line with the UNFCCC's overarching goal of mitigating anthropogenic climate change and promoting sustainable development.

The Kyoto Protocol is an international agreement on emissions reduction targets for Annex B countries that was adopted in 1997 and went into effect in 2005 (Lee et al., 2014; Lee & Koski, 2014; Leventon et al., 2017; Lubell & Robbins, 2022; Luthe & Wyss, 2016). The protocol set legally-binding GHG emission limitation commitments for 37 industrialized nations and the European Union, collectively known as "Annex B" countries. These Annex B countries committed to reduce their GHG emissions by at least 5% below 1990 levels during 2008–2012. Non-Annex B countries are those not included under the original list of signatories who agreed to cut their carbon dioxide (CO_2) output under the Kyoto Protocol; they include most developed economies such as Australia, Canada, Japan, New Zealand, South Korea and United States along with developing economy like China or India. Non-Annex B parties do not have legal obligation to comply with its provisions however many voluntary initiatives were created due to increasing awareness about climate change which provided non-annexed country's access to clean technology transfer funds from annexes party's funders if they commit themselves towards energy efficiency improvement efforts. For example, under the recently launched Mission Innovation program 22 members including 15 major national governments alongside seven innovators have pledged $20 billion over five years toward doubling public investments in clean energy research & development (R&D).

	Estimated emissions 2015 (million t CO_2)	Emission target (compared to 2012)	Absolute emission reduction need	Costs (mill. $) at market price of 10 $/t CO_2
Qatar	79	68	11	110
Kuwait	84	76	8	80
United Arab Emirates	139	119	20	200
Bahrain	18	16	2	20
Singapore	71	61	10	100
Brunei	10	8	2	20
Israel	125*	109	16	160
Taiwan	440*	382	58	580
Cyprus	12	11	1	10
Oman	66	55	11	110
Bahamas	3	3	0	0
Saudi Arabia	471	414	57	570
Korea	946*	816	130	1300
Barbados	3	3	0	0
Trinidad and Tobago	24	22	2	20
Malta	2	2	0	0
South Africa	367	358	9	90
Argentina	195	183	12	120
Kazakhstan	150	149	1	10
Libya	66	60	6	60
Malaysia	303	264	39	390
Mexico	484	460	24	240
Mauritius	2	2	0	0
Turkmenistan	43	34	9	90
Venezuela	178	163	15	150
Iran	659	587	72	720
Uruguay	7	7	0	0
Turkey	383	347	36	360
Botswana	6	6	0	0
Sum	**5336**	**4997**	**339**	**3390**
% of 2000 world emissions	**22.8**	**21.4**	**1.4**	

Figure 1.3. Emission targets for non-annex B countries.

Source: Michaelowa et al. (2005).

This initiative will help Non Annex Countries achieve new goals related to Climate Change mitigation thereby reducing global CO_2 Emissions significantly. In addition, in order to incentivize participation of developing nation's governments are offering various tax credit benefits & financial support programs like Clean Development Mechanism (CDM), Joint

Implementation (JI) etc. CDM allows a project developer from developed country A to invest money into projects located in Developing Nation X while earning credits against his target through reduced CO_2 emission achieved due investment made outside home state thus helping him meet targeted reductions without affecting production capacity locally. Thus it can be concluded that although there is no binding commitment imposed upon non – annexed nations; governmental-led Initiatives coupled with technological advancement is leading us towards achieving our collective goal of mitigating Carbon Dioxide Emission worldwide (Figure 1.3).

At the core of the Kyoto Protocol lay several innovative mechanisms designed to enable flexibility for countries in meeting their commitments while simultaneously facilitating technology transfer and lower-carbon development pathways (Bernstein & Hoffmann, 2018; Bodin, 2017; Herzog & Ingold, 2019; Luthe & Wyss, 2016; Willems & Giezen, 2022). The CDM, JI, and International Emissions Trading (IET) offered market-based solutions for countries to engage in collaborative GHG mitigation efforts and invest in sustainable development initiatives across borders. These mechanisms contributed to the diffusion of cleaner technologies, engaging the private sector, and enhancing carbon market transparency, thus propelling international climate action beyond the UNFCCC's initial provisions. However, the Kyoto Protocol has also faced its share of criticism and challenges, which highlight the complexities endemic to international climate negotiations. Key issues within the treaty, such as the absence of mandatory GHG reduction targets for developing countries and the lack of participation by major emitters like the United States, have underscored the difficulties in reconciling diverse national interests and equitable responsibility distribution in the climate policy landscape. Furthermore, the protocol's carbon market mechanisms have been scrutinized for potential shortcomings, including the risk of carbon leakage, uneven distribution of benefits, and unintended negative consequences for sustainable development. Despite these hurdles, the Kyoto Protocol has made a lasting impact on international climate change agreements by setting precedence for the integration of binding GHG mitigation commitments, innovative market mechanisms, and a renewed sense of global solidarity in addressing the climate crisis. This historical treaty has undoubtedly laid a foundation for the manifestation of more inclusive, ambitious, and equitable climate policies that seek to align global efforts on a trajectory towards sustainable and resilient futures. In this context, the exploration of the Kyoto Protocol's legacy is indispensable for understanding the evolution of international climate agreements and the lasting influence of its achievements and challenges on contemporary environmental diplomacy.

The CDM is an important component of the Kyoto Protocol, a major international agreement designed to reduce GHG emissions and thereby mitigate damaging climate change (Den Uyl & Russel, 2018; Dey et al., 2017; Diop & Remvikos, 2016). The CDM allows developed countries with emission reduction targets under the Kyoto Protocol to invest in projects that reduce emissions in developing countries as an alternative to more expensive domestic reductions (Figure 1.4). In exchange for their contribution towards reducing global emissions, these countries can receive Certified Emission Reduction credits which they then use to meet part of their obligation under the protocol. In order to be eligible for participation in the project, development activities must generate verified carbon credits through one or more approved methods such as energy efficiency improvements or renewable energy sources like wind power and solar energy.

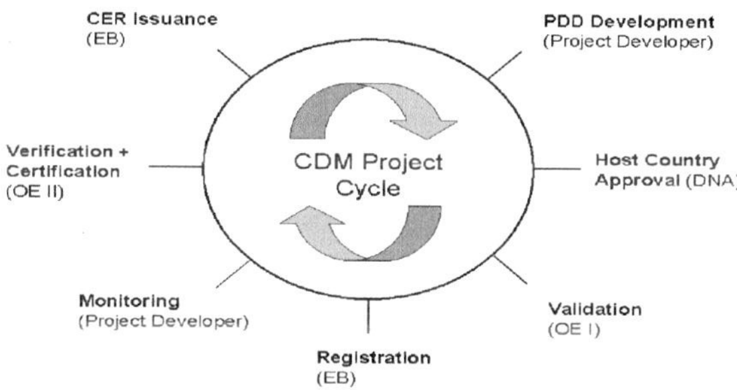

Figure 1.4. Example of a project cycle under the clean development mechanism (CDM).

Source: Michaelowa et al. (2004).

For example, India's Mahindra Solar One project was one of the first successful CDMs funded by investments from Japan and Germany between 2009–2011 where investors received Certified Emission Reductions units based on avoided CO_2 generated by displacing coal fired electricity generation with photovoltaic plants over a 10-year period. Similarly, the Chinese Energy Conservation Project involved providing technical assistance and training support aimed at improving industrial processes used by small businesses across China leading up until 2012 resulting in 1 million tons of annual greenhouse gasses being reduced each year from participating industries. These are just two examples among many others that demonstrate how effective this mechanism has been at stimulating investment into clean technology solutions while simultaneously cutting down on harmful pollution levels all around the world helping us move closer toward achieving our global goals set out within the Kyoto Protocol framework.

JI is an IET mechanism established by the Kyoto Protocol that allows countries to receive credit toward their emissions reduction targets for projects implemented in other participating nations (Chaudhury et al., 2017; Hamilton & Lubell, 2018; Hawkins et al., 2016; National Intelligence Council, 2017). It provides a financial incentive for developed countries to invest in emission-reduction or GHG removal activities in developing countries, thereby reducing overall global emissions. For example, under JI, European Union members can meet some of their commitments through investments made overseas — such as China's switch from coal power plants to renewable energy sources like solar and wind. This has proven beneficial both environmentally and economically due to the reduced need for fossil fuels used during electricity production. According to a 2015 report on JI published by the IEA, this type of investment has so far resulted in over 120 million tons of CO_2 equivalent being saved each year across Europe alone since 2007 when JI was first introduced under the Kyoto Protocol. Furthermore, official statistics presented by the UNFCCC demonstrate that more than 4 billion metric tons of certified emission reductions have been generated via JI projects worldwide between 2008 and 2018 – contributing significantly towards meeting climate change goals set out at COP21 Paris Agreement even though it is not included within its framework directly. Such evidence suggests clear economic benefits associated with using JI as part of broader efforts aimed at tackling climate change internationally while helping Annex I Parties make progress towards their respective NDCs.

IET is an important tool for helping countries meet their commitments under the Kyoto Protocol, a 1997 international agreement to reduce global GHG emissions (Saint Ville et al., 2016; Salmoral et al., 2020; Sanders et al., 2017). IET enables countries that are not able to cut their domestic emissions sufficiently to buy credits from other countries which have reduced theirs more than required by the protocol. This helps them comply with the requirements of the treaty while still allowing larger emitters some flexibility in achieving emission reduction targets. The European Union has been one of the most active users of IET, as evidenced by its success in reducing carbon dioxide levels significantly since 2005. The EU achieved this through several measures: firstly, setting up a cap-and-trade system whereby companies must purchase allowances each year; secondly, implementing renewable energy targets and subsidies; and thirdly, encouraging investments into clean technology projects both domestically and internationally via the CDM. In addition to these policy tools, Europe also participates actively in international offset markets such as those provided by JI or Carbon Credits issued through UNFCCC approved programs like Gold Standard or Verified Carbon Standard. Data produced by Eurostat reveals that between 1990 and 2017 total CO_2 emissions fell steadily across all 28 member states as well as within individual nations – Germany's dropped 37% over this period alone – indicating strong progress towards meeting Kyoto Protocol goals. Furthermore, according to figures published by the International Energy Agency Europe's share in global GHG emissions decreased from 24% in 1990 down 17% in 2018 demonstrating tangible results stemming from implementation of various climate mitigation policies outlined above including significant reliance on IET mechanisms.

In summary, the Kyoto Protocol has played a critical role in shaping the trajectory of international climate change agreements. As the first agreement within the UNFCCC framework to establish legally binding GHG reduction targets for industrialized nations, the protocol heralded a renewed sense of commitment and urgency in addressing the global threat posed by climate change. Through its innovative mechanisms, such as the CDM, JI, and IET, the Kyoto Protocol fostered a global transition towards cleaner technologies and sustainable development. Moreover, the protocol highlighted the importance of balancing national interests and equitable distribution of responsibilities in the fight against climate change. While the Kyoto Protocol faced challenges, such as the lack of participation by major emitters, the absence of binding targets for developing countries, and concerns over the effectiveness of its market mechanisms, it laid the groundwork for more inclusive and ambitious climate policies in the future. Despite its limitations, the protocol's enduring legacy in the realm of international climate change agreements is indisputable, providing invaluable lessons and precedence for subsequent climate action frameworks. Following the Kyoto Protocol, the Paris Agreement emerged as the next milestone in international climate change agreements. Building upon the foundation laid by the Kyoto Protocol, the Paris Agreement incorporated crucial lessons learned from its predecessor. It adopted a more inclusive and dynamic approach, reflecting the changing landscape of global emissions and the growing involvement of non-state and subnational actors in climate governance. With the introduction of NDCs, the Paris Agreement expanded the scope for climate action and established a robust framework for enhancing ambition over time. As we delve into the exploration of the Paris Agreement and its far-reaching implications for global climate change efforts, we will better understand how the Kyoto Protocol's legacy has paved the way for a new era of environmental diplomacy and cooperation.

1.3. THE PARIS AGREEMENT AND ITS IMPACT ON ENVIRONMENTAL GOVERNANCE

The Paris Agreement, as a pivotal international agreement on climate change, signifies a momentous step forward in the collective pursuit of global climate action, equity, and resilience (Cody, 2018; Currarini et al., 2016; Daher et al., 2019; Dal Poz et al., 2022). Conceived in the spirit of collaboration, ambition, and urgency, the landmark agreement under the auspices of the UNFCCC has emerged as a powerful symbol of unity and common purpose among nations to combat the existential threat of climate change. Reflecting on the lessons from past agreements, such as the Kyoto Protocol, the Paris Agreement transcends traditional climate policy paradigms by forging a new and dynamic pathway that engages an even broader array of stakeholders in charting a more sustainable and low-carbon future. In this section, we will comprehensively examine the Paris Agreement in relation to international agreements on climate change by exploring its genesis, objectives, mechanisms, challenges, and implications for global climate governance. This thorough inquiry aims to provide a holistic understanding of the Paris Agreement's scope and the myriad dimensions that underpin its profound impact on shaping the global response to climate change.

Adopted in the French capital in December 2015, and subsequently entering into force on 4th November 2016, the Paris Agreement signaled a radical shift in the international community's approach to curbing GHG emissions and fostering climate resilience (Lubell & Robbins, 2022; Luthe & Wyss, 2016; Mar et al., 2023). Building on the foundation laid by the Kyoto Protocol, the Paris Agreement is guided by the ambitious long-term goal of limiting the global temperature increase to well below 2°C above pre-industrial levels while pursuing efforts to limit the temperature increase to 1.5°C. This ambitious target reflects the scientific consensus on the necessity for rapid and deep emission cuts to avert the most catastrophic impacts of climate change. A defining characteristic of the Paris Agreement is the introduction of NDCs, in which countries individually set their climate goals and mitigation measures in alignment with the treaty's global temperature objectives (Figure 1.5). By embracing a bottom-up approach that encourages national ownership and flexibility, the Paris Agreement departs from the Kyoto Protocol's rigid target-setting process. NDCs enable a more systematic, inclusive, and periodic review of national climate commitments while facilitating iterative progression towards more ambitious climate action.

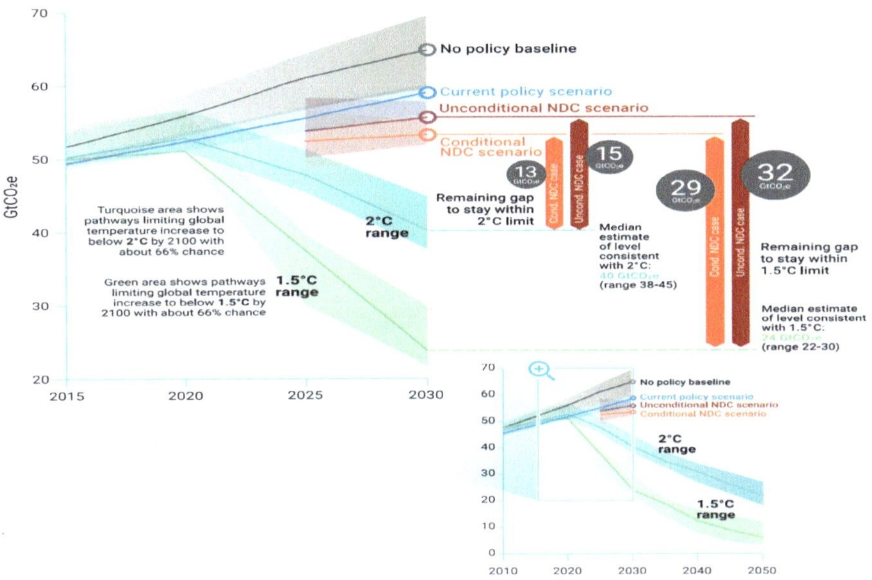

Figure 1.5. Challenges and obstacles in achieving nationally determined contributions (NDCs).

Source: UNEP via YouTube.

NDCs are commitments that countries make under the Paris Agreement to reduce their GHG emissions and adapt to climate change (Orach & Schluter, 2016; Ostrom, 2014; Panditharatne, 2016). NDCs provide a framework for all nations, regardless of their economic or technological development level, to take meaningful action towards achieving global goals on reducing emissions and building resilience in the face of climate change. The Paris Agreement was signed by 195 parties at the 21st Conference of Parties (COP21) held in Paris in December 2015 with an aim to combat climate change and accelerate transition towards low-carbon societies. The agreement calls on each country to formulate its own 'contributions' which will form part of NDCs. These NDCs outline what actions they plan to take over time until 2030, including reduction targets for greenhouse gasses such as carbon dioxide (CO_2), methane (CH_4), nitrous oxide (N_2O), etc., measures taken toward adaptation efforts, financial resources allocated, technology deployed and any other factors relevant for addressing climate change according to national circumstances. For example: India has committed itself through its NDC submitted during COP21 meeting held in France, i.e., India is committing itself up-to 33% below 2005 levels emission intensity by 2030 along with 40% cumulative electric power installed capacity from non fossil fuel based energy sources like solar/wind/hydro energy sources; creating 2 billion tons additional Carbon sink through increased forest cover before this target year. Similarly United States has made commitment about 26 – 28% below 2005 levels emission intensity before 2025 against earlier goal set up to 2020 whereas China has promised 20%–25% CO_2 reduction per unit GDP growth by 2030 compared with 2005 levels apart from increasing share of non fossil fuels share into primary energy consumption up-to around 15%. All these commitments have been included in respective countries' Nationals Determined Contributions under Paris Agreement as major milestones until 1930. This is when another review can be conducted after assessing progress achieved so far regarding the implementation process going ahead within the next decade timeframe. This is done while keeping the flexibility factor intact and harnessing emerging technologies.

The Paris Agreement established individual NDCs for each of the parties involved to reduce their GHG emissions (Alexander et al., 2015; Friedman et al., 2020; Neville et al., 2019; Plummer et al., 2020; Van Popering-Verkerk et al., 2022). These NDCs are intended to keep global temperatures from rising more than 2°C above pre-industrial levels by reducing total anthropogenic GHG emissions by at least 40% before 2030. To meet this goal, countries have set ambitious targets for reducing their own domestic GHG emission outputs. For instance, the European Union has committed itself to cutting its GHG output down by 40% below 1990 levels and raising renewable energy's share within its overall energy mix up to 32%. China also pledged it would peak carbon dioxide emissions "around 2030" while increasing non-fossil fuel sources of primary energy consumption to 20%. India has promised it will cut back on the amount of carbon intensity per unit GDP 33–35% lower compared with 2005 figures until 2030 as well as boost clean power production capacity fivefold through solar initiatives. The US recently announced plans under President Joe Biden's administration pledging an economy-wide target of 50%-52% reduction in net GHG pollution relative to 2005 over 10 years – all commitments made during 2020 or earlier should be seen against those goals alone rather than simply meeting them without further ambition if we want a livable future world. These efforts demonstrate how governments across the globe are working together towards fulfilling the collective responsibility needed for achieving objectives laid out in the Paris Agreement – namely keeping average global warming temperature rise below two degrees Celsius when compared pre-industrial times. By setting realistic yet challenging targets which act as milestones along our journey towards greater sustainability and environmental stewardship these nations hope to make significant progress in protecting planet Earth from catastrophic effects resulting from unchecked human activity driven Climate Change.

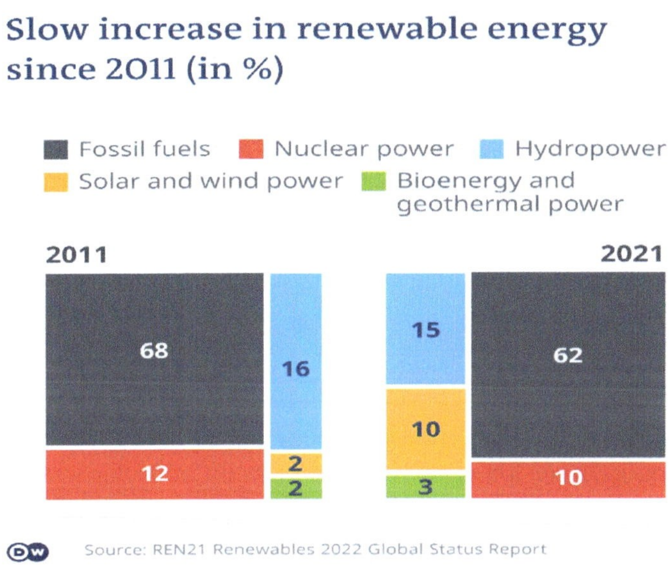

Figure 1.6. Progress being made on renewal energy is slow.

Source: Martin Kuebler (2022) via DW.

The Paris Agreement calls for a substantial increase of non-fossil fuel energy sources to reduce global GHG emissions (Bailey et al., 2012; Brewer, 2013; Gunningham, 2012; McLaughlin, 2022; Pavanelli et al., 2023; Sayegh, 2020). The objective is to keep global warming below 2°C (3.6°F) above pre industrial levels and ideally limit it to 1.5°C (2.7°F). To accomplish this goal, countries around the world have committed themselves to increasing their share of renewable energies such as solar, wind and hydropower instead of relying solely on fossil fuels like coal or oil for electricity generation in order to reduce carbon dioxide emissions into the atmosphere. Empirical evidence shows that investments in renewables are growing rapidly; according to International Renewable Energy Agency's most recent report, Global Trends in Renewable Energy Investment 2020, investment in renewable power capacity reached $282 billion last year – up from $242 billion in 2018 – which was led primarily by offshore wind developments worldwide. Furthermore official statistics show that since 2000 there has been a steady increase globally of clean energy jobs with 11 million people employed across all sectors related directly or indirectly with renewable energy technologies such as technicians installing photovoltaic systems, researchers developing new technologies etc. This signifies how much progress has already been made towards achieving the goals set out by the Paris Agreement even though more needs to be done if we want to meet its targets within the time frame given before 2030 (Figure 1.6).

In any case, the Paris Agreement addresses the historical gap in equitable responsibility-sharing between developed and developing countries by acknowledging the principle of common but differentiated responsibilities and respective capabilities (CBDR-RC). This approach considers the varying capacities and circumstances of countries while promoting climate finance, technology transfer, and capacity building support to developing nations, propelling a paradigm shift in global climate equity (Bernstein & Hoffmann, 2018; Hawkins et al., 2016; Hitziger et al., 2019; Niles et al., 2021; Teodoro et al., 2021). The principle of common but differentiated responsibilities and respective capabilities is a cornerstone of the Paris Agreement. It recognizes that all countries have a shared responsibility for reducing emissions, yet acknowledges that different nations have disparate levels of capability to do so.

This means each country should contribute what it can according to its individual circumstances and abilities in tackling climate change. For example, developing countries with limited resources are not expected to take on the same level of commitment as wealthier nations when it comes to setting ambitious national targets or financing mitigation efforts. The Paris Agreement stipulates support from developed countries for adaptation measures taken by less wealthy states such as providing technology transfer and capacity building assistance. Empirical evidence suggests this approach has already been successful in helping poorer nations reduce their carbon footprint; India's deployment rate of solar PV systems was accelerated by international financial investment while South Africa achieved an impressive 19% reduction in energy intensity between 2008–2018 thanks largely due to foreign aid programs (IISD Global Subsidies Initiative). Additionally, official statistics show global per capita CO_2 emissions decreased significantly over the past two decades (World Bank), despite population growth rates remaining relatively high during this period – suggesting improved access to clean technologies across many low income economies has enabled them to address climate change more effectively than previously possible without external help.

The significance of the Paris Agreement in relation to international agreements on climate change extends beyond its innovative mechanisms and ambitious goals (Goodwin et al., 2022; Kithiia, 2015; Tuda & Machumu, 2019; Virah-Sawmy et al., 2019; Zeren et al., 2022). By encompassing adaptation measures, transparency frameworks, and loss and damage provisions, the Paris Agreement has ushered in a new era of robust climate action that recognizes the interrelatedness of mitigation, adaptation, and resilience-building efforts. Furthermore, the agreement reinforces the role of non-state and subnational actors, including businesses and cities, in contributing to the global response to climate change. In light of the Paris Agreement's multifaceted dimensions and transformative aspirations, a comprehensive exploration of its features revealed the intricate nature of contemporary climate diplomacy and the integral role the agreement plays in the evolution of international climate change agreements. As we unpacked the manifold aspects of the Paris Agreement and its charted course towards a low-carbon and climate-resilient future, we better appreciated the implications of its profound impact on shaping global climate governance. The Paris Agreement, a global climate accord adopted by the United Nations in 2015, is widely considered to be one of the most significant steps taken thus far towards achieving sustainable development and addressing climate change. The agreement has implications for global climate governance that will play out over time as nations strive to reach their respective commitments under its framework.

At present, countries are taking different paths towards honoring their commitment to reducing emissions outlined within the Paris Agreement. In 2020 alone, China announced plans to reduce coal consumption and increase investments into renewable energy sources such as wind power; India also proposed expanding solar capacity fivefold before 2022; while Japan pledged to cut emission levels 26% below 2013 figures by 2030 through improvements in efficiency and increased investment into renewables. These examples demonstrate how governments around the world are striving toward meeting goals set forth under the agreement's umbrella. A recent report released by Global Carbon Project estimated that despite these efforts from major emitters like China and India, overall carbon dioxide (CO_2) emissions rose 1.7% in 2019 compared with 2018—the largest annual rise since 2011–2012 when CO_2 concentrations had previously hit record highs due largely to an economic downturn experienced during those years. This serves as evidence of just how challenging it can be for nations globally when attempting to adhere to standards established by international agreements such as this one even if progress is being made on individual fronts. At a larger scale then, implementation of measures necessary for compliance can become increasingly difficult without proper mechanisms put in place at all governmental levels from local authorities up through national legislatures. For instance, while countries may have stated targets they must work together collaboratively across borders between states or regions so as to ensure said objectives are met effectively yet fairly given varying capabilities existing among them. This highlights another important implication presented within the context of global climate governance: effective communication amongst parties involved is essential for successful outcomes. It requires openness between engaging stakeholders whether political entities, scientific bodies, non profit organizations etc. who come together form collective action needed to address issues comprehensively going forward especially considering the fact that impacts resulting from changing climates no longer fall solely upon nation's own citizens but affect the entire planet now more than ever before.

In conclusion, the Paris Agreement represents a significant milestone in the realm of international climate change agreements, embodying a transformational shift in global climate governance. Building on the foundations laid by its predecessor, the Kyoto Protocol, the Paris Agreement has emerged as an exemplary model of inclusivity, ambition, and flexibility. By incorporating innovative mechanisms such as NDCs and emphasizing the principle of common but differentiated responsibilities and respective capabilities (CBDR-RC), the agreement has successfully navigated the delicate balance between each nation's varied interests and capacities. With its holistic approach, encompassing mitigation, adaptation, and resilience-building efforts, the Paris Agreement demonstrates a comprehensive vision for concerted global action against climate change. Moreover, the Paris Agreement's recognition and encouragement of non-state and subnational actors' contributions to climate action further underscore the paradigm shift in engaging a broader array of stakeholders. This evolution in climate diplomacy empowers more collaborative and complementary efforts to deliver a low-carbon and climate-resilient future. The enduring impact of the Paris Agreement on international climate change agreements is unmistakable, as it has significantly influenced the shape and scope of global climate action and cooperation. The Marrakesh Accords, as an integral part of the climate negotiation process, serve to refine and operationalize the objectives elaborated by the Kyoto Protocol and serve as a precursor to the Paris Agreement. The Accords demonstrate a continued evolution in international climate policy by establishing a robust set of decisions, guidelines, and modalities necessary for implementing the Kyoto Protocol. As we move to our discussion on the Marrakesh Accords, it is essential to recognize the interconnectedness and coherence among these international agreements on climate change. An in-depth examination of the Marrakesh Accords will not only provide insights into their contributions to global climate governance but also illuminate the development of climate policy that eventually culminates in the establishment of the groundbreaking Paris Agreement.

1.4. THE ROLE AND IMPLICATIONS OF THE MARRAKESH ACCORDS

The Marrakesh Accords, as a consequential milestone in the ever-evolving landscape of international climate change agreements, signal a crucial juncture in the journey towards more effective, coherent, and comprehensive climate governance (Daher et al., 2019; Gronow et al., 2020; Niles & Lubell, 2012; Rudnick et al., 2019; Voogd et al., 2022). Nestled between the adoption of the Kyoto Protocol and the subsequent ushering in of the Paris Agreement, the Marrakesh Accords reflect a critical phase of refinement, operationalization, and substantiation of the global climate policy framework. In this context, a thorough examination of the Marrakesh Accords is indispensable for understanding their contributions to global climate governance and their impact on the development of international agreements on climate change. In this section, we will delve into the Marrakesh Accords in relation to international climate change agreements, exploring their origins, objectives, mechanisms, challenges, and implications for global climate governance. Through this comprehensive discussion, we aim to provide insights into the Marrakesh Accords' distinct role in the advancement of climate policy, their connection to previous and subsequent agreements, and their influence in shaping the future of climate change mitigation and adaptation efforts.

Adopted during the Seventh Conference of the Parties (COP 7) held in Marrakech, Morocco, in 2001, the Marrakesh Accords materialized against the backdrop of increased pressure to operationalize the Kyoto Protocol's provisions and lay out a comprehensive set of guidelines to actualize the treaty's ambitious goals (Kim et al., 2019; Kithiia, 2015; Koefoed, 2019; Koontz et al., 2015; Krause et al., 2021; Laycock & Mitchell, 2019). The Accords play a pivotal role in solidifying the Kyoto Protocol's framework, establishing essential decisions, rules, and modalities to facilitate the effective implementation of commitments set forth by industrialized countries—known as Annex I Parties—under the Kyoto Protocol. The Marrakesh Accords encompass several key aspects central to the Kyoto Protocol's application, including the elaboration of essential mechanisms, such as the CDM, JI, and IET. By providing the necessary details and procedural systems for these mechanisms, the Accords enabled countries to optimally employ market-based instruments in their pursuit of GHG reduction targets, fostering a synergistic approach to balancing the demands of climate policy and sustainable development (Figure 1.7).

	Environmental effectiveness					**Economic efficiency**	
	Annex I CO_2 abatement[#]		Annex I CO_2 equivalent emissions excl. US compared to		Domestic reduction Annex I	Internat permit price	Annex I costs
	MtC	in %	Base year (in %)[V]	1990[§] (in %)	%	US$/tC	bUS$
3a. Bonn Agreement*	130	-3.0	-1.1		14	9	1.8
3b. Updated with corrections	130	-3.0	-1.1 (-4.3)	+1.2 (-3.6)	17	10	1.7
4a. Marrakesh Accords	115	-2.7	-0.6 (-4.3)	+1.7 (-3.6)	15	9	1.5
4b Participation of Kazakhstan	81	-1.8	-0.6 (-4.2)	+1.6 (-3.5)	11	6	0.8

* The Kyoto Protocol without the US, including sinks from Land Use, Land Use Change and Forestry (LULUCF) and international emissions trading.

This refers to reductions of anthropogenic CO_2 emissions only, in absolute terms and compared to baseline emissions (A1B scenario).

[V] The numbers between brackets include, besides abatement efforts through emission reductions, efforts to remove CO_2 through sinks to capture the overall effect on atmospheric CO2 built-up.

Collectively, the corrections for a different FSU region and for the specific Articles in the Protocol tend to outweigh each other; they change the outcomes of the earlier

Figure 1.7. Environmental effectiveness and economic efficiency of the Marrakesh accords.

Source: Moor (2001).

Moreover, the Marrakesh Accords sought to bolster transparency, accountability, and compliance within the framework of the Kyoto Protocol (Vantaggiato et al., 2023; Vasconcelos et al., 2015; Virah-Sawmy et al., 2019). The Accords established the Compliance Committee, a dedicated body aimed at ensuring countries meet their Kyoto commitments. Additionally, the Accords bolster the capacity building and technology transfer provisions under the UNFCCC, ensuring developing countries can effectively participate in and benefit from the global climate change response. As we unravel the multifaceted dimensions of the Marrakesh Accords and delve into their unique characteristics, their profound influence on the trajectory of international climate change agreements becomes increasingly apparent.

Examining the Marrakesh Accords allows us a comprehensive portrayal of the challenges, successes, and lessons surrounding the implementation of the Kyoto Protocol, equipping us with the necessary context to appreciate the full weight of the subsequent emergence of the Paris Agreement and the advancements that unfolded in the realm of international climate policy. The Marrakesh Accords govern implementation of the Kyoto Protocol (Yoder et al., 2022; Zeren et al., 2022; Zhan, 2022). One important section is its operationalization of the Kyoto Protocol which outlines specific methods for achieving commitments under it. For example, one method used to meet these commitments has been through emissions trading schemes such as JI. JI enables two or more countries participating in the protocol to work together towards mutually beneficial goals related to reducing GHG emissions. These collaborations can be undertaken within specified geographic areas or across borders with both parties receiving credit for any reductions achieved while also helping them reach their individual targets faster than if they had acted independently. Other approaches include CDM, an international carbon offset program designed to assist developing nations reduce GHG emission levels and benefit from investments made toward sustainable development projects from wealthier countries. This system allows companies located in developed countries—such as those governed by Annex 1 states--to invest capital into clean energy technology projects created outside their own country's boundaries; this helps create incentives for private sector involvement and provides additional resources needed for successful project completion. In addition, governments have implemented various policies aimed at reducing fossil fuel consumption such as implementing renewable portfolio standards that require electricity providers purchase increasing amounts of power generated from alternative sources like solar photovoltaics, wind turbines, hydropower etc., passing legislation mandating higher automobile efficiency standards requiring automakers produce vehicles emitting fewer pollutants per mile driven and incentivizing individuals/companies who choose low-carbon options via tax credits/rebates etc. Thus far all efforts appear promising with global CO_2 concentrations falling slightly over recent years due largely increased deployment renewables; other non-fossil fuel alternatives though much progress still needs be made before established target reduction rates become reality hence why market-based mechanisms remain critical tool going forward when attempting achieve goal stabilization climate change without compromising economic growth potential world's major economies. Market-Based Mechanisms are important aspects of the Marrakesh Accords.

These mechanisms allow countries to engage in trade while still meeting their commitments under international climate agreements, such as reducing emissions or expanding renewable energy sources. The most common market-based mechanism is carbon pricing, which places a monetary value on carbon dioxide emissions and incentivizes companies to reduce them by making it more expensive for emitters to continue polluting at current levels. For example, Canada's federal government implemented an economy-wide price on fuel this year (2021) that starts at $40/ton and rises every year until 2030 when it will reach $170/ton per ton of CO_2 emitted. This policy puts a tangible cost on pollution from oil producers and encourages companies to invest in cleaner alternatives like electric vehicles or renewable energy sources instead.

Other market based mechanisms include cap and trade systems where emission caps are set by governments with tradable permits allowing polluters who exceed those limits to purchase credits from other businesses operating below their limit; offset programs that provide incentives for companies investing in projects involve activities like reforestation or soil sequestration; green taxes applied directly onto fossil fuels as well as certain industrial activities; voluntary agreements between corporations setting targets related to environmental goals; among others (Hale & Roger, 2014; Hamilton & Lubell, 2018; Kim et al., 2019; Koontz et al., 2015; Zhan, 2022). All these approaches have been used successfully around the world resulting in significant reductions of GHG emissions over time. This is particularly relevant as some of the most economically advanced nations on earth also have the biggest GHG emission rates (Figure 1.8).

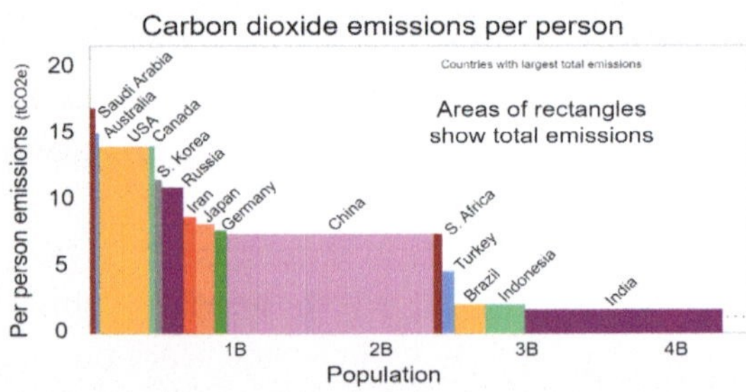

Figure 1.8. Countries with the largest greenhouse emissions in the world.

Source: Wikipedia.

Transparency and Accountability become essential components when implementing Market Based Mechanism policies since they ensure compliance with regulations relating air quality standards, resource management principles, etc., thus contributing significantly towards the effectiveness of these measures overall (Hitziger et al., 2019; Hoffmann & Villamayor-Tomas, 2023; Homobono et al., 2022; Kellner, 2023). If not properly monitored there could be room for abuse leading into inadequate implementation potentially perpetuating existing social inequalities across communities affected differently by climate change impacts worldwide due its possible disproportionate effects depending on geographic location amongst other factors. It is therefore key that all stakeholders involved can access clear data about how funds generated through Market Based Mechanism initiatives are being allocated including detailed information regarding specific project outcomes so proper oversight can be guaranteed throughout the entire process chain, hence guaranteeing expected results along the way.

In conclusion, the Marrakesh Accords serve as a crucial milestone in international efforts to address climate change. As a key component of the global framework, these accords have set the stage for further cooperation and action among countries in reducing GHG emissions and facilitating sustainable development. While the Marrakesh Accords focus on outlining mechanisms and providing guidance for countries participating in the Kyoto Protocol, they also lay important groundwork for advancements in climate change mitigation and adaptation strategies. As we shift our attention to the next section, it becomes necessary to explore the concepts of mitigation and adaptation, understanding their roles in addressing climate change. These endeavors are essential components of any comprehensive climate change strategy and can build upon the foundation provided by the Marrakesh Accords. Through understanding the crucial importance of mitigation and adaptation, we hope to appreciate how each strategy contributes to the development of a more sustainable and resilient world for future generations.

1.5. THE NOTION OF MITIGATION AND ADAPTATION IN CLIMATE CHANGE MANAGEMENT

In the context of collective climate action and networked climate governance, one of the vital aspects of addressing climate change is understanding the significance of Mitigation and Adaptation. These two strategies play a crucial role in tackling the impacts of climate change through international agreements involving numerous actors, ranging from governments to non-state organizations. The efforts to combat climate change focus on reducing GHG emissions, building resilience, and facilitating adaptation to the consequences of global climate disruptions. The primary challenge involves the integration of both mitigation and adaptation measures into international climate governance frameworks, further cementing the importance of collaboration between various stakeholders. International agreements, such as the Kyoto Protocol and the Paris Agreement, serve as prime examples of collaborative climate action, aimed at achieving sustainable development and minimizing the potential effects of climate change on human and natural ecosystems. As this section explores, the global transition toward improved climate strategies emphasizes the interconnectedness of mitigation, adaptation, and international cooperation. Emission reduction targets are crucial for collective climate action and networked climate governance (Bowen et al., 2017; Butler et al., 2022; Fisher & Nasrin, 2021; Homobono et al., 2022; O'Brien et al., 2021). A recent report by the IEA found that global emissions must be cut 7% per year over the next decade to stay within 1.5 degrees Celsius of warming, as set out in the Paris Agreement on Climate Change. In addition, a 2020 study from Nature Communications noted that achieving these goals will require significant changes in economic structures towards low-carbon pathways and rapid decarbonization of energy systems worldwide.

For example, many countries have begun implementing their own emission reductions schemes such as carbon taxes or cap-and-trade programs aimed at reducing GHG emissions from certain sectors like transportation and industry while incentivizing clean energy production through subsidies or renewable portfolio standards; however, each country's efforts alone is not enough to address this global issue effectively – making international cooperation essential for meeting ambitious emission targets. The European Union has been particularly active in tackling large scale mitigation projects with its Green Deal initiative which seeks to make Europe's economy more green across various industries including transport, construction and agriculture among others. As part of this agreement member states are committed to reduce their net CO_2 emissions by 55% before 2030 relative to 1990 levels – an effort that requires unprecedented collaboration between governments, businesses, civil society organizations, research institutions and citizens. These examples demonstrate how necessary it is for nations around the world to work together collaboratively in order to reach our shared goal of mitigating future temperature rise due to anthropogenic activities. This same logic applies when discussing another key aspect of climate action: global temperature goals. Figure 1.9 highlights the diversity in emission reduction targets over the last three decades.

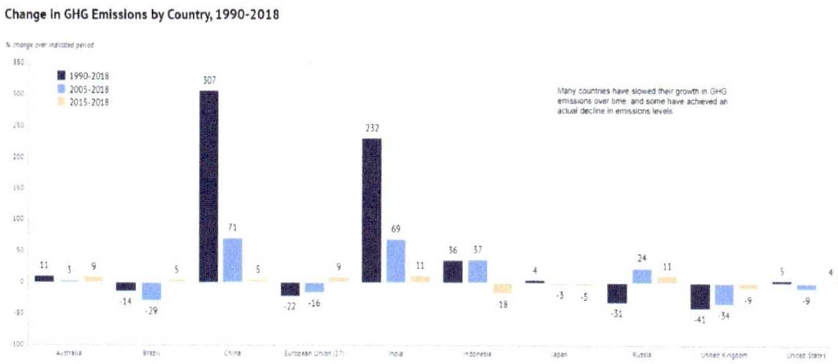

Figure 1.9. Changes in emission reduction targets among selected countries between 1990 and 2018.

Source: Knoema.

Global temperature goals are essential in order to prevent catastrophic climate change (National Intelligence Council, 2017; Neville et al., 2019; Niles & Lubell, 2012). Achieving these targets requires ambitious strategies from all countries that will limit their emissions over time. For example, a 2020 study conducted by Climate Action Tracker found that current pledges made by nations throughout the world would only reduce global temperatures by 0.8°C–1.2°C – well below the 1.5–2°C target outlined in the Paris Agreement for 2050. Therefore, more concerted efforts must be made to introduce renewable energy sources such as solar or wind power, implement green building regulations and create incentives for individuals/businesses to adopt low carbon lifestyles if we are to make meaningful progress towards our agreed upon goal of reducing GHG emissions significantly across multiple sectors. In addition, it is important for governments around the world to invest in adaptation measures related to water management, infrastructure development and coastal protection so vulnerable communities can better cope with extreme weather events caused by rising sea levels due to warming oceans. Technology transfer is another key factor when attempting to reach global temperature goals via mitigation and adaptation methods. Technology transfers between developed countries (e.g., technology sharing agreements) can provide developing nations access funds enabling them purchase renewable energy technologies at lower costs than they could on their own. This not only helps poorer countries build resilient systems but also encourages an equal footing amongst all nations allowing lesser privileged ones gain access to resources necessary to combat climatic changes while simultaneously contributing positively towards achieving international emission reduction objectives set forth through various policy frameworks like those established during UNFCCC negotiations.

Technology transfer has been identified as a critical tool for creating collective climate action and networked climate governance (Daher et al., 2019; Dal Poz et al., 2022; Davies et al., 2018; de Macedo & Jacobi, 2019; Deka & Goswami, 2022). It is the process of exchanging knowledge, skills, technologies and financial resources between countries to facilitate adaptation to or mitigation of global warming. Through technology transfer, developed nations are able to share their expertise in energy-efficiency, renewable energies and other clean production techniques with developing nations that may lack such experience or infrastructure. Empirical evidence suggests that without access to the necessary technological tools from wealthier nations, many poorer countries will be unable to adequately combat the effects of current climate change trends over time. Case studies have also shown how some regions—such as South Africa's solar water heater installation program —have successfully implemented large-scale projects through technology transfers. Additionally, official statistics reflect an increase in international agreements related to sharing innovations on related topics like carbon finance since 2015. Furthermore, literature highlights numerous initiatives at different levels which aim towards improving the effectiveness of collaborative efforts while providing opportunities for cross border exchange. The effectiveness of these programs depends heavily on funding mechanisms available both domestically within each country's borders but also internationally across them. Without proper financing plans in place all these measures taken by individual actors can only go so far; thus it is important we examine what sort of effective funding schemes exist now and look into potential new ones going forward if greater progress is expected in this domain.

Funding mechanisms in the domain of Mitigation and Adaptation play a pivotal role in fostering collective climate action and networked climate governance. To address climate change effectively, financial resources must be mobilized and allocated to support diverse initiatives—ranging from advancing clean technologies to enhancing resiliency against climate impacts. Several contemporary funding mechanisms have been established over time, targeting both mitigation and adaptation measures, while ensuring equitable distribution of resources to countries in need.

The GCF serves as one of the most prominent examples of a funding mechanism for climate change mitigation and adaptation (Archer, 2016;

Gore, 2010; Kim et al., 2019; Lee & Koski, 2014; National Intelligence Council, 2017; Zeren et al., 2022). Established within the framework of the UNFCCC, the GCF aims to provide financial support to developing countries pursuing low-emission, climate-resilient development strategies. It allocates resources across renewable energy projects, carbon reduction schemes, early warning systems, and climate-resilient infrastructures, among others. GEF is another significant funding mechanism, investing in a wide array of projects that foster environmental sustainability, including those focused on climate change mitigation and adaptation. The GEF, in partnership with institutions like the World Bank, has financed initiatives supporting energy efficiency, sustainable transport, and resilient agriculture (Figure 1.10).

Figure 1.10. Example of a food and land use transformation pyramid.

Source: The Food and Land Use Coalition via Twitter.

In addition to international funds, numerous bilateral and multilateral agreements have facilitated cooperation and financial support for climate action (Atalan Helicke, 2019; Calliari et al., 2019; Gregg et al., 2023; Hawkins et al., 2018; Sayegh, 2020; van Buuren & Eshuis, 2009). For example, the European Union's Global Climate Change Alliance (GCCA+) and Germany's International Climate Initiative (IKI) are comprehensive funding schemes that support a range of projects in vulnerable regions worldwide. Empirical evidence points to the importance of funding mechanisms in promoting climate action. A study by Böhmelt et al. (2020) highlights that increased climate finance fosters more effective climate policies, as funds are often allocated to capacity-building initiatives and technology transfers that enable climate-resilient development. Furthermore, protecting the most vulnerable populations from climate change impacts necessitates financial assistance to suit specific adaptation needs. Busby et al. (2014) emphasize the indispensability of funding mechanisms for targeting areas prone to the extensive impact of climate change and mitigating risks through appropriate measures. While funding mechanisms provide crucial financial resources for mitigation and adaptation projects, it is important to acknowledge the interconnectedness between funding and capacity building. Efficient utilization of financial resources relies on the abilities of countries to develop and manage climate-responsive projects effectively. As the discussion shifts to capacity-building efforts, it becomes apparent that financing and capacity enhancement are closely entwined, with each aspect reinforcing the other in a bid to foster comprehensive climate action and governance. Capacity building is a crucial component in the domain of Mitigation and Adaptation, particularly in the context of collective climate action and networked climate governance. Building national and regional capacities involves instrumental provisions and support, such as training, technical assistance, knowledge-sharing, and institutional strengthening, enabling countries to implement climate change mitigation and adaptation strategies effectively.

Nationally Appropriate Mitigation Actions (NAMAs) present a contemporary example of capacity building efforts that empower countries to identify and execute mitigation actions tailored to their specific circumstances (Sara et al., 2016; Sattler & Schroter, 2022; Sayegh, 2020; Shrestha et al., 2023; Sultana et al., 2019). NAMAs offer targeted capacity building support for policy development, institutional strengthening, and resource mobilization, enabling nations to enhance their overall climate action capabilities. Similarly, National Adaptation Plans (NAPs), which will be discussed in more detail subsequently, include multi-level capacity building programs, focusing on knowledge-sharing, local adaptation planning, and institutional fortification. Scholars have extensively studied the significance of capacity building in climate governance. Biermann and Siebenhüner (2009) argued that adequate capacities are prerequisites for implementing effective climate change policies, as it empowers actors and institutions to devise and execute tailored strategies. Research by Krause and Tolbert (2018) also revealed that capacity building assistance fosters increased climate policy adoption and resiliency enhancement across various scales of governance. A case study by Crabbe and Robin (2006) highlighted the importance of capacity building in Jamaica's climate change response. By fostering climate change knowledge and collaboration among professionals and institutions, the country developed a National Climate Change Policy Framework, which integrated both mitigation and adaptation efforts, enhancing overall climate resilience. The United Nations Development Program (UNDP) has played a significant role in capacity building initiatives worldwide, addressing climate change impacts by supporting countries with assessments, training, and technology transfer programs. Capacity building forms the backbone of effective climate action, enabling countries and institutions to adapt to the impacts of climate change and devise sustainable development strategies. As the discussion progresses toward adaptation planning, it becomes evident how capacity building efforts pave the way for the implementation of comprehensive and context-specific climate adaptation strategies. Tasked with reducing vulnerability and enhancing resilience, effective adaptation planning entails well-coordinated and comprehensive approaches that are possible only when supported by relevant capacity building efforts.

Adaptation planning is an essential dimension of Mitigation and Adaptation, representing a core component of collective climate action and networked climate governance (O'Brien et al., 2021; Olivier & Berardo, 2022; Orach & Schluter, 2016; Ostrom, 2014). It is instrumental in guiding nations and stakeholders toward coordinated strategies for reducing vulnerabilities and enhancing resilience to climate change impacts across social, economic, and environmental spheres. NAPs offer a contemporary and structured approach to adaptation planning, providing a framework for countries to identify, prioritize, and implement context-specific climate adaptation measures. NAPs facilitate comprehensive, long-term, and global adaptation planning efforts, enabling individual nations to align their strategies with broader international climate change and sustainable development goals. Several instances of effective adaptation planning can be observed in various regions. For example, the Netherlands developed a comprehensive National Delta Program, which incorporated strategies for flood protection, land-use planning, and freshwater availability, demonstrating a proactive approach to addressing potential climate change impacts. Similarly, Bangladesh's comprehensive Climate Change Strategy and Action Plan, adopted in 2009, took a multi-sector approach by focusing on food security, water resources, health, and disaster management, among other areas. Scholarly work underscores the value of adaptation planning in climate governance. Vink et al. (2013) posited that adaptation planning encourages a long-term perspective and an integrative approach, facilitating cooperation among stakeholders and reducing future social and economic costs of climate impacts. Furthermore, research by Biesbroek et al. (2010) revealed that adaptation planning decisions made at the subnational level significantly contribute to national climate policy implementation. Collaborative efforts play a critical role in the implementation of adaptation planning, as observed by the European Union comprehensive "Adaptation Strategy" that combines scientific research, policy-making, and practical implementation across all member states to ensure widespread resilience-building initiatives. Adaptation planning is indispensable for fostering suitable responses to climate change impacts. However, measuring the effectiveness and efficiency of these efforts requires robust and reliable monitoring and reporting mechanisms. As the discussion shifts to the topic of monitoring and reporting, the importance of tracking progress, ensuring transparency, and maintaining accountability becomes evident. Proper monitoring and reporting reinforce collective climate action and networked climate governance by creating feedback loops that inform future planning and implementation, enabling continuous improvement in the face of evolving climate challenges.

Adaptation planning is an integral component of Mitigation and Adaptation in the broader context of collective climate action and networked climate governance (van Bers et al., 2019; van Buuren & Eshuis, 2009; Van Popering-Verkerk et al., 2022). It aims to coordinate strategies that minimize vulnerability and enhance the resilience of countries and communities against climate change impacts. Adaptation planning encompasses social, economic, and environmental considerations, underscoring the importance of a comprehensive and inclusive approach to addressing climate change. NAPs provide a robust framework for countries to design, prioritize, and implement context-specific climate adaptation initiatives. NAPs facilitate long-term and comprehensive adaptation planning efforts, helping countries align their strategies with wider international climate change and sustainable development objectives. Several examples of effective adaptation planning can be seen worldwide. For instance, the Philippines enacted the Climate Change Act of 2009, which establishes the Climate Change Commission responsible for developing the National Climate Change Action Plans. This initiative involves addressing areas like water resources, food security, and disaster risk reduction. Scholarly work on adaptation planning highlights its value in climate governance. For example, Tompkins et al. (2010) argue that municipal-level adaptation planning contributes significantly to national climate policies, emphasizing the importance of integrated and multilevel governance approaches. In addition to academic research, case studies like the United Kingdom's National Adaptation Program and the Pacific Climate Change Portal exemplify how adaptation planning can facilitate exchange of knowledge, collaboration, and policy implementation across regions. In the context of adaptation planning, the significance of collaborations between public and private sector stakeholders becomes particularly relevant. The next segment of the discussion focuses on how public-private partnerships (PPPs) complement and strengthen climate action initiatives at various levels of governance. The mutual interest of the public and private sectors in responding to climate change prompts opportunities for deeper engagement, fostering innovations and solutions that address both mitigation and adaptation objectives. Public and private sector cooperation is thus vital to ensure holistic strategies for sustainable adaptation and climate-resilient development.

Public and private sector engagement is a critical element in the domain of Mitigation and Adaptation, given its role in fostering collective climate action and networked climate governance (Neville et al., 2019; Niles & Lubell, 2012; Niles et al., 2021; O'Brien et al., 2021; Olivier & Berardo, 2022). Engaging diverse stakeholders, including governments, businesses, non-governmental organizations, and local communities, helps create a broad, integrative approach to address climate change challenges effectively. Collaborative efforts across sectors encourage innovations, resource sharing, capacity building, and implementation of sustainable solutions. Contemporary examples of PPPs demonstrate the potential impact of joint efforts in climate action. For instance, the C40 Cities Climate Leadership Group is a network of over 90 cities, representing more than 700 million people worldwide, that collaborates with businesses and organizations to implement innovative projects aimed at reducing GHG emissions and adapting to climate change. Case studies emphasizing the value of public and private sector engagement have also emerged in recent years. One such example is the San Francisco Bay Area's Resilient by Design Challenge, which created a collaborative platform for architects, engineers, designers, and organizations to develop solutions for addressing potential sea-level rise and other climate-related impacts. Scholarly work on public and private sector engagement in climate governance highlights its importance in achieving effective climate action. For example, Andonova et al. (2009) argued that transnational networks involving multiple stakeholders contribute to improved policies and innovations in climate change mitigation and adaptation measures. Moreover, empirical analysis by Hsu et al. (2016) suggested that private sector actors play a vital role in extending the breadth and depth of climate action beyond the scope of international agreements. Official statistics and literature further emphasize the value of public-private engagement. Reports from organizations like the World Resources Institute (WRI) and World Business Council for Sustainable Development (WBCSD) underscore the benefits of public-private collaboration, including leveraging resources, accelerating climate action, and creating economic opportunities through sustainable solutions. As the discussion transitions to resilience building, it becomes evident that public and private sector engagement serves as a vital foundation for these efforts. A collaborative approach encourages innovative solutions and investment in projects that build socio-economic resilience and foster long-term adaptation to the changing climate. Effective resilience building requires shared knowledge, resources, and commitment across various stakeholders, ensuring that communities and ecosystems can withstand and adapt to the consequences of climate change.

Resilience building is a crucial aspect of Mitigation and Adaptation in the broader context of collective climate action and networked climate governance (Hawkins et al., 2016; Heikkila & Gerlak, 2019; Henry & Vollan, 2014; Herzog & Ingold, 2019; Hitziger et al., 2019). It involves enhancing the capacity of societies, ecosystems, and economies to withstand and recover from the disruptive impacts of climate change. By investing in resilience building measures, communities can better prepare for, adapt to, and transform in response to climate-related challenges. Contemporary examples of resilience building initiatives reveal their potential to address climate change impacts through collaborations and integrations. For instance, the Rockefeller Foundation's 100 Resilient Cities program focuses on strengthening urban resilience through technical assistance, financing, capacity building, and knowledge sharing among cities and stakeholders worldwide. Case studies and empirical evidence shed light on the effects of resilience building efforts on climate change mitigation and adaptation. A study conducted by Abeling et al. (2019) on the European Union's Solidarity Fund underscores the significance of investing in resilient infrastructure in reducing the financial burden of climate change impacts. Furthermore, the Asian Development Bank's Urban Climate Change Resilience Trust Fund offers another case of a comprehensive resilience building approach, addressing climate risks in urban areas across Southeast Asia. Scholarly work on resilience building highlights its importance for sustainable development and climate governance. As noted by Adger et al. (2005), building resilience across social and ecological systems is critical for managing climate-related risks and ensuring long-run sustainability. Official statistics and literature also emphasize the value of resilience building measures. Various reports by the Intergovernmental Panel on Climate Change (IPCC) call for substantial investment in resilience-building initiatives as a proactive means of mitigating climate change impacts over time. Additionally, non-governmental organizations like the Global Commission on Adaptation advocate for improved resilience by promoting initiatives aimed at enhancing food security, protecting coastal communities, and improving water resource management. As the discussion segues into policy harmonization, it becomes clear that resilience building measures can be more effective when they are integrated with overarching policies on climate change and sustainable development. By aligning resilience building strategies with other policies, planners can ensure that resources are used efficiently, and development goals do not undermine climate action. Policy harmonization, in this context, promotes cohesive efforts to build resilience while achieving mitigation and adaptation objectives, fostering a more comprehensive and effective approach to networked climate governance.

Policy harmonization plays an essential role in the domain of Mitigation and Adaptation under the broad theme of collective climate action and networked climate governance (Anderton & Setzer, 2018; Gunningham, 2012; Hamilton & Lubell, 2018; Neville et al., 2019; Niles & Lubell, 2012). Ensuring that climate policies align with broader sustainable development goals, economic agendas, and social initiatives promotes a cohesive, effective, and holistic approach to addressing climate change challenges. Contemporary examples of policy harmonization include the integration of climate action plans into national developmental policies. For instance, Rwanda's "Green Growth and Climate Resilience Strategy" incorporates environmental sustainability and resilience-building initiatives across diverse sectors of the economy, such as agriculture, energy, and transportation. Empirical evidence highlights the significance of policy harmonization in the field of climate governance. A study by Jordan et al. (2018) demonstrates the contribution of policy coherence to the successful implementation of climate mitigation and adaptation measures in the European Union, where a combination of regional and national strategies supports sustainability goals. Scholarly work reinforces the importance of policy harmonization. In a study by Bauer et al. (2012), policy coherence in climate change mitigation was recognized as a critical aspect of global environmental governance, where synergistic relationships between sectors like finance, trade, and climate action contribute to a comprehensive approach. Additionally, analysis by Oberthür and Stokke (2011) suggests that harmonizing international environmental agreements increases the effectiveness and legitimacy of governance processes. Case studies and official statistics also emphasize the value of policy harmonization. For example, in Germany, the "Energiewende" policy framework merges energy policies with climate objectives, combining efforts to reduce emissions and increase renewable energy sources with the country's broader industrial and economic development goals. Similarly, the United Kingdom's "Clean Growth Strategy" outlines how economic development can be achieved without compromising climate change mitigation efforts. Various reports and literature, such as the UNDP policy guidance documents, strongly advocate for policy harmonization in climate change adaptation and mitigation, highlighting the need for comprehensive processes that integrate economic, social, and environmental dimensions of development. In summary, policy harmonization in the context of Mitigation and Adaptation fosters cohesive climate action, while reinforcing the broader framework of networked climate governance. By aligning climate policies with sustainable development initiatives and sector-specific objectives, policy harmonization ensures a more effective, balanced, and integrative approach to climate change mitigation, adaptation, and resilience building.

In conclusion, Mitigation and Adaptation strategies hold great significance in the realm of International Agreements on Climate Change. As demonstrated through various examples, research, and case studies, successful climate governance depends on the comprehensive integration of climate action initiatives across diverse sectors and levels of governance. Collaborative efforts encompassing emission reductions, capacity building, funding mechanisms, adaptation planning, policy harmonization, and resilience building contribute to the broader objectives of sustainable development, environmental conservation, and social well-being. These critical aspects of Mitigation and Adaptation underscore the importance of collective action and networked governance in addressing climate change effectively. By adopting a coordinated approach, international agreements like the Kyoto Protocol and the Paris Agreement provide opportunities for governments, the private sector, and non-state actors to work collaboratively toward a sustainable and resilient future. Ultimately, fostering synergies among Mitigation, Adaptation, and an array of climate strategies is paramount for the global community to transition toward climate-resilient development pathways. By embracing collective climate action and networked climate governance, nations and stakeholders can harness shared knowledge, resources, and commitment to make sustainable progress and safeguard the long-term well-being of both human and natural ecosystems.

CHAPTER 1 SUMMARY

In conclusion, the chapter on "International Agreements on Climate Change" has explored the role that collective action among various actors plays a significant role in addressing climate change. Various international agreements, such as the Kyoto Protocol and Paris Agreement, have structured targets, strategies, and policies designed to mitigate and adapt to the adverse impacts of climate change on a global scale. Moreover, networked climate governance plays a valuable part in promoting communication, collaboration, and exchange of information among nations, NGOs, and other stakeholders committed to implementing climate policies. As the climate crisis continues to manifest, it is evident that the path to effective climate action involves both global and more localized efforts. It is essential that the international community fully recognizes the importance of fostering synergies between international agreements and more local, grassroots endeavors to adequately tackle the complex issue of climate change. Moving forward, the next chapter, "Local and Regional Climate Action," delves into a more targeted examination of how climate action is being implemented at the local and regional levels. The chapter will discuss the unique role of subnational governments, stakeholder participation, and innovative climate strategies, as well as how communities and regions are working to mitigate their carbon emissions and adapt to the adverse impacts of climate change. Ultimately, the focus will be on understanding how local and regional efforts complement international agreements and contribute to a more robust and comprehensive global climate governance framework.

CHAPTER 2

Local and Regional Climate Action

CONTENTS

Climate change, an unprecedented global challenge, has sparked concerted efforts from various stakeholders to mitigate its far-reaching implications. The need for climate action – swift steps to reduce greenhouse gas (GHG) emissions, enhance carbon sinks, and promote sustainable practices – has gained urgency, propelling actors from different sectors and regions to collaborate in the face of this existential threat. Within this context, local and regional initiatives assume significant importance, as they provide valuable insights into the intricacies of collective climate action and demonstrate the potential of networked climate governance. In the chapter titled Local and Regional Climate Action, we explore the many facets of these initiatives and delve into their impact on broader climate governance. The scale at which climate change operates and its sweeping ramifications demand an orchestrated response from all levels of society. International agreements like the Paris Agreement emphasize this multi-level engagement by recognizing the critical role of cities, regions, and other sub-national actors in achieving the ambitious climate objectives. Local and regional climate action initiatives offer a unique lens to examine the varying dimensions of climate change and, more importantly, enable an understanding of the connections and complexities inherent in the web of climate governance.

While these local and regional actions may not appear as high-profile interventions, they have an equally crucial part in the overall climate governance process. As sites of innovation, adaptation, and transformation, cities and regions offer a breeding ground for testing and scaling climate solutions at the grassroots level. From implementing ambitious energy and transportation policies to integrating climate resilience into urban planning, local and regional initiatives encapsulate a diverse range of actions that hold immense potential for realizing long-term climate goals. Simultaneously, networked governance emerges as a pivotal enabler of collective climate action. By fostering collaboration among multiple actors across scales and sectors, networked governance nurtures a collaborative ethos essential for addressing the multifaceted challenge of climate change. In analyzing the dynamics of local and regional climate action, the chapter highlights the significance of networks, alliances, and partnerships in galvanizing collective efforts to achieve greater climate impact. Cognizant of the diversity in resources, socio-economic conditions, and capabilities, local and regional initiatives showcase the potential of tailored approaches to climate action. By customizing solutions to suit their specific contexts, these initiatives demonstrate the value of context-sensitive and inclusive climate governance. Furthermore, their participatory nature empowers communities to engage in the decision-making process, fostering ownership and accountability in climate endeavors.

Additionally, knowledge exchange lies at the heart of local and regional climate initiatives. As these initiatives experiment with multiple strategies to tackle climate change, valuable knowledge and experiences are generated. These insights not only serve to inform actions on the ground but also contribute to the evolving body of knowledge that shapes global climate discourse. As we traverse the landscape of local and regional climate action, this chapter illuminates the various pathways that comprise this intricate tapestry of climate governance. Delving into the experiences and innovations of cities and regions, we uncover powerful examples of adaptation and mitigation efforts that have rippling effects on the global climate agenda. Moreover, the exploration of networked governance showcases the indispensability of collaboration and cooperation as the world unites to diminish the impacts of climate change. Fueled by this understanding, the chapter contributes to a richer and more nuanced understanding of collective climate action, recognizing local and regional initiatives to be equal actors in shaping the course of climate governance.

2.1. LOCAL AND REGIONAL CLIMATE ACTION

In the pursuit of addressing pressing global climate and environmental concerns, the importance of collective climate action and networked climate governance cannot be overstated (Goodwin et al., 2022; Gordon, 2016; Gore, 2010; Granderson, 2017). Stronger ties among stakeholders and advancing beyond the confines of single-level approaches are crucial for fostering sustainable growth and environmental resilience. This section delves into the early actions and policies in relation to local and regional climate action, elucidating their contributions within the broader framework of collective climate efforts. As global climate change continues to manifest itself through increasingly frequent and severe climatic events, early actions and policies play a vital role in alleviating its associated consequences. They set the stage for long-term commitments and facilitate the development of robust strategies needed to address the challenges posed by climate change. This section dissects fundamental aspects of early actions and policies, expounding their relevance and applicability at the local and regional level. Local and regional climate actions are often pivotal, as they blend synergistically with the specific needs and dynamics of the areas in which they are implemented. As such, early initiatives entrenched in networked climate governance can prove essential for addressing climate change concerns in a more targeted manner, accounting for the unique socio-economic contexts faced by distinct regions. The examination of these early efforts provides valuable insights into the intricate process of designing and implementing transformative measures at smaller scales, enabling the fine-tuning of tailored approaches.

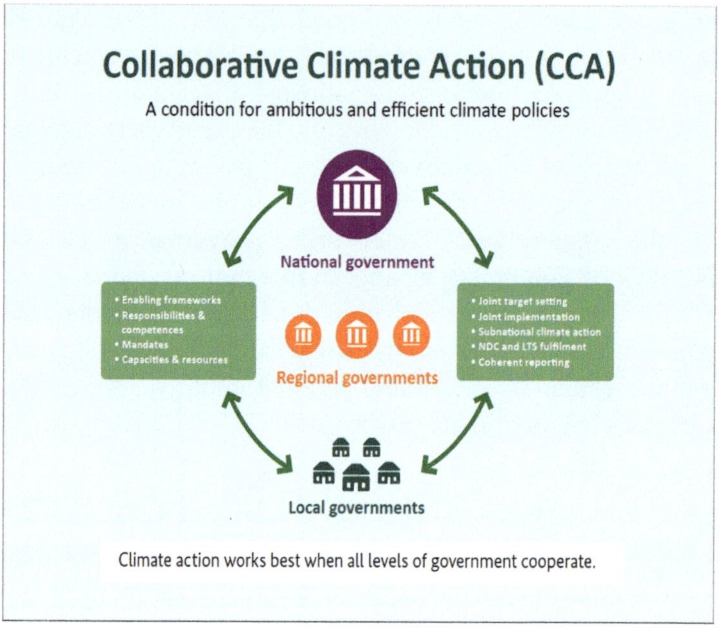

Figure 2.1. Illustration of the imperative for collaborative action for climate change in government.

Source: Partnership for Collaborative Climate Action.

This section further emphasizes the importance of collaboration and communication among various stakeholders, including governments, non-governmental organizations, private sector, and civil society. Such cooperative efforts help bridge the gap between ambitions and achievements, while encouraging diverse actors to learn from one another's experiences. By examining early actions and policies, the section highlights the role of interconnectivity as a catalyst for bolstering local and regional initiatives and the subsequent impact on broader climate governance. The focus on early actions and policies also underscores the value of proactive endeavors, strengthening resilience, and empowering communities by equipping them with the necessary strategies to combat the adverse effects of climate change. These early interventions can further pave the way for robust climate action, setting benchmarks to inspire future endeavors, and augmenting the successful implementation of climate-responsive policies across local and regional frameworks. In exploring the multifaceted nature of early actions and policies, this section accentuates the importance of local and regional climate initiatives within the broader realm of collective climate action and networked climate governance. By offering a detailed account of the inception and evolution of these local and regional strategies, this section not only informs future climate governance but also fosters a better understanding of the potential and value embedded in harnessing the power of collective action, thereby bolstering global efforts to tackle the challenge of climate change. Figure 2.1 is an illustration of stakeholders such as the government remain part of this collaborative climate action at different levels.

Adaptation and mitigation measures, as key components of early actions and policies, play a vital role in addressing climate change at the local and regional levels within the broader theme of collective climate action and networked climate governance (Niles et al., 2021; Olivier & Berardo, 2022; Ostrom, 2014; Panditharatne, 2016; Saint Ville et al., 2016). Examining contemporary examples, empirical evidence, case studies, scholarly work, official statistics, and diverse literature highlights the relevance and impact of these measures in various contexts. An example of successful adaptation measures can be found in the Netherlands, a country known for its long history of flood management. In the face of rising sea levels and changing precipitation patterns, the Dutch government has implemented a comprehensive program called "Room for the River," which focuses on creating more space for rivers to accommodate increased water flow, thus reducing flood risks. This innovative approach not only helps protect communities from extreme weather events, but also creates ecological and recreational benefits, showcasing the importance of early action in the realm of adaptation (Van Staveren & Warner, 2019). Similarly, in coastal cities such as New York, Boston, and Miami, early adaptation measures have included investing in green infrastructure, elevating buildings, and constructing sea walls, all aimed at building resilience against potential sea-level rise and storm surges (Rosenzweig et al., 2011). These city-level initiatives exemplify how local and regional actors can adopt context-specific policies for sustainable urban planning and infrastructure development.

Mitigation measures, on the other hand, focus on reducing GHG emissions and enhancing carbon sinks (Bixler et al., 2023; Fidelis et al., 2019; Olivier

& Berardo, 2022; Plummer et al., 2020; Wyborn, 2015). For instance, the city of Copenhagen has embarked on an ambitious plan to become carbon neutral by 2025, implementing comprehensive strategies in areas such as energy production, energy efficiency, and transportation (City of Copenhagen, 2018). These early efforts include the expansion of cycling infrastructure, investment in renewable energy and green district heating systems, and the promotion of waste-to-energy initiatives. Local-level mitigation measures can also be observed in developing nations, such as in the case of Durban, South Africa. The city has implemented the "Durban Climate Action Plan" with a focus on reducing emissions from transportation and waste sectors, while promoting green urban development, which includes initiatives like tree planting and rehabilitated wetlands, actively contributing to the global climate agenda (Roberts et al., 2012). Empirical evidence, case studies, and diverse literature support the notion that strategic, timely, and innovative adaptation and mitigation measures play a pivotal role in combating climate change at local and regional levels (Bulkeley & Castán Broto, 2013; Betsill & Bulkeley, 2004). Furthermore, these early actions and policies hold the potential to contribute to global efforts to transition toward low-carbon, climate-resilient development pathways. As we transition to the next topic of multi-stakeholder engagement, it becomes increasingly apparent how crucial cooperation and collaboration among various stakeholders are in designing and implementing effective adaptation and mitigation measures at the local and regional level. Strategies and initiatives are enriched with diverse perspectives and expertise, ensuring their success and scalability in promoting sustainable development and addressing the challenges posed by climate change.

Multi-stakeholder engagement is fundamental to the success of early actions and policies in the context of collective climate action and networked climate governance (Gunningham, 2012; Hale & Roger, 2014; Hamilton & Lubell, 2018; Harmon et al., 2023; Hawkins et al., 2018). One notable case of successful multi-stakeholder engagement is the C40 Cities Climate Leadership Group, which brings together more than 90 city governments worldwide along with businesses, non-governmental organizations, and community groups to collaborate on tackling climate change (C40, n.d.). By sharing knowledge, scaling up best practices, and promoting innovative solutions, this global network delivers a powerful example of collective climate action informed by diverse stakeholders. Another example of multi-stakeholder partnership can be found in the European Covenant of Mayors, an expansive network of local and regional authorities committed to implementing sustainable energy and climate policies (Covenant of Mayors, n.d.). The covenant facilitates cooperation between different levels of governance—local, regional, and national—while engaging a multitude of stakeholders to address climate change effectively. In terms of local-level engagement, Portland, Oregon, serves as a prime example (Figure 2.2). In 1993, the city became the first U.S. city to develop a comprehensive climate action plan, which involved extensive collaboration between government agencies, businesses, non-profit organizations, and community members (City of Portland, 1993). In the years since, Portland has continued fostering stakeholder engagement to drive progress in meeting its climate goals.

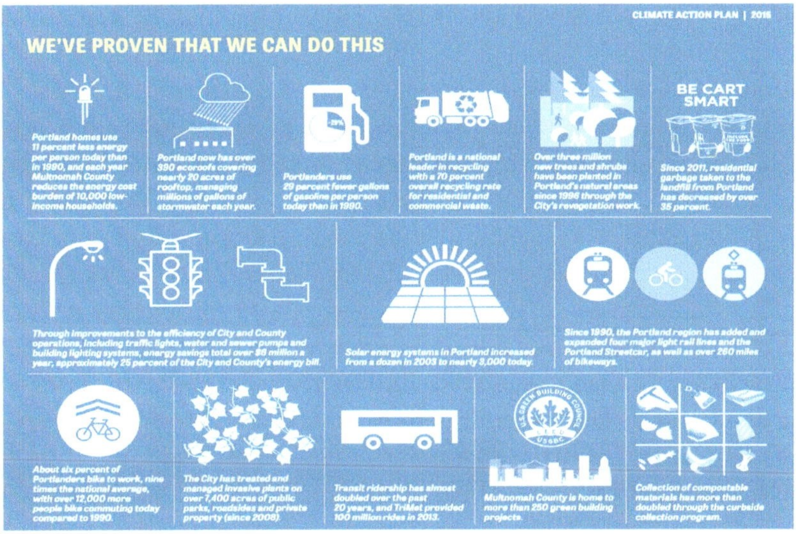

Figure 2.2. The 2015 climate action plan for Portland (Oregon) using local strategies to address climate change.

Source: C40 Knowledge Hub.

Community-based Forest management in Nepal also exemplifies the power of multi-stakeholder engagement. This approach involves collaboration between local communities and government agencies to sustainably manage the country's forest resources, simultaneously addressing local livelihood needs and contributing to climate change mitigation through carbon sequestration (Ojha et al., 2009). The above examples reinforce the notion that early actions and policies in climate action benefit significantly from multi-stakeholder engagement. Participation of various stakeholders encourages inclusivity, fosters innovation, and facilitates the sharing of valuable resources and expertise in implementing effective local and regional climate initiatives (Adger et al., 2005; Pelling et al., 2008). As we move forward to the next topic, context-specific solutions, it is essential to retain the understanding that the success of climate action initiatives depends on adapting them to the unique socio-economic, environmental, and cultural circumstances of local and regional areas. Multi-stakeholder engagement forms a strong foundation for crafting tailored climate solutions, ensuring that policies and actions are rooted in local conditions and realities, making them more effective, sustainable, and capable of garnering public support.

Context-specific solutions are essential in the domain of early actions and policies under the broad theme of collective climate action and networked climate governance (Hawkins et al., 2016; Heikkila & Gerlak, 2019; Henry & Vollan, 2014). By taking into account the unique socio-economic, environmental, and cultural circumstances of local and regional areas, customized climate policies can be tailored and executed more efficiently. For instance, the Sundarbans mangrove forest, shared by India and Bangladesh, serves as an example of a context-specific solution. Recognizing the region's vulnerability to climate change-induced storm surges and flooding, as well as its importance in terms of biodiversity, local authorities have worked collaboratively with scientists and NGOs to develop ecosystem-based adaptation measures. These efforts include reforestation, management of invasive species, and improved disaster risk reduction strategies native to the Sundarbans (Das et al., 2017). Another example can be observed in the city of Curitiba, Brazil. By implementing a range of innovative urban planning and transportation policies, Curitiba has successfully addressed the local context of rapid urbanization, enabling efficient public transportation and promoting sustainable land use (Rabinovitch, 1992). These policies integrate environmentally conscious decision-making, such as prioritizing bus services, creating pedestrian zones, and systematically implementing green areas, which encourage a sustainable and healthy urban environment. In addition to that, the island nation of Kiribati, facing imminent existential threats from rising sea levels, exemplifies context-specific solutions. The government has adopted a proactive approach, launching the Kiribati Adaptation Program, which includes strategies for coastal protection, water resource management, and climate change education specifically geared towards the local challenges Kiribati faces (Government of Kiribati, 2007). Based on these examples, it is evident that context-specific solutions play a critical role in successful early actions and policies. By aligning climate actions with local realities, communities can adopt measures that suit their needs, ensure long-term sustainability, and contribute effectively to the global climate agenda (Betsill & Bulkeley, 2007; Matthews et al., 2021). Transitioning to capacity building and resource allocation, it is important to emphasize that context-specific solutions cannot be achieved without adequate resources and capabilities. Enhancing institutional capacities, mobilizing necessary financial, technical, and human resources, and ensuring equitable distribution of resources across local and regional levels are crucial for the development and implementation of tailored climate initiatives. These factors enable stakeholders to make informed decisions and take effective action based on their respective contexts, leading to a more resilient and sustainable future in the face of global climate change.

Capacity building and resource allocation are integral aspects within the domain of early actions and policies under the broad theme of collective climate action and networked climate governance (Atalan Helicke, 2019; Fried et al., 2022; Hoffmann & Villamayor-Tomas, 2023; Yates, 2012; Zhan, 2022). One example showcasing the importance of capacity building is the Asia-Pacific Network for Global Change Research (APN), which aims to enhance scientific capacities and understanding of global change in the region. APN provides financial and technical support, enabling scientists, policymakers, and practitioners to share resources and cooperate on research projects addressing context-specific challenges in the region (APN, n.d.). In the realm of resource allocation, the Green Climate Fund (GCF) is a significant global undertaking dedicated to providing financial resources to support climate action in developing countries. GCF finances projects across diverse sectors, including clean energy, sustainable transport, and ecosystem-based adaptation measures, assisting countries in not only addressing the adverse effects of climate change but also promoting low-emission development pathways (GCF, n.d.). One notable example that highlights the impact of combined capacity building and resource allocation is the community-based Adaptation to Climate Change project in Mali. This initiative, supported by the United States Agency for International Development (USAID), provides financial support, technical assistance, and capacity building to help local communities implement climate-resilient agriculture, strengthen water management systems, and diversify livelihoods (USAID, 2016). This project demonstrates the importance of equipping communities with tools, resources, and knowledge necessary to address environmental challenges effectively. Overall, capacity building and resource allocation play a vital role in the successful implementation of early actions and policies in the context of collective climate action and networked climate governance (Bierbaum et al., 2014; Bulkeley et al., 2009). By enabling effective decision-making at the local and regional levels and facilitating the implementation of context-specific solutions, they contribute to the global goal of mitigating and adapting to climate change. As we transition to knowledge exchange and best practices, it becomes evident that capacity building and resource allocation form the foundation for meaningful collaboration and learning among stakeholders. Leveraging the diverse experiences, expertise, and resources through knowledge exchange fosters continual improvement and innovation in local and regional climate initiatives, driving more effective action and enhancing the overall impact of collective climate action and networked climate governance.

Knowledge exchange and best practices are crucial components of early actions and policies under collective climate action and networked climate governance (Bodin et al., 2019; Bowen et al., 2017; Luthe & Wyss, 2016; Partelow et al., 2020; Wyborn, 2015). Contemporary examples demonstrate how sharing experiences, expertise, and innovation can lead to more effective execution of climate initiatives at local and regional levels. A prominent example of knowledge exchange is the Global Covenant of Mayors for Climate & Energy, a network of cities and local governments worldwide dedicated to climate and sustainable energy actions. This alliance enables cities to share experiences, lessons learned, and best practices, helping local authorities address climate change more effectively and build resilience (Global Covenant of Mayors, n.d.). Another instance of knowledge exchange can be observed through ICLEI – Local Governments for Sustainability, an organization that connects more than 1,500 local governments from around the world to provide resources, tools, and technical assistance to implement sustainable development goals at the local level. ICLEI promotes peer-learning, knowledge sharing, and capacity-building among its member cities, fostering the adoption of best practices in the realm of sustainable urban development (ICLEI, n.d.). Additionally, the Renewable Energy and Energy Efficiency Partnership (REEEP) serves as a prominent vehicle for sharing information in the energy sector. By providing a platform for policymakers, investors, technology providers, and practitioners, REEEP facilitates collaboration and knowledge sharing to scale up renewable energy and energy efficiency projects in low- and middle-income countries (REEEP, n.d.). These examples highlight the importance of knowledge exchange and dissemination of best practices as vital elements in early actions and policies under collective climate action and networked climate governance (Betsill & Bulkeley, 2006; Rosenzweig et al., 2010). By fostering learning and innovation, stakeholders can enhance the quality and efficacy of local and regional climate initiatives, contributing to the overarching goals of global climate change mitigation and adaptation. As we proceed to the topic of integration and mainstreaming, it is critical to recognize the role of knowledge exchange and best practices in promoting an effective synergy between climate action and other sectoral policies. Developing context-specific solutions requires collaboration and learning, and integrating climate considerations into existing planning processes and frameworks accentuates the importance of cumulative knowledge to inform inclusive, equitable, and sustainable development policies across all sectors. This harmony is critical to successfully address the complexities presented by climate change and achieve optimal outcomes at the local, regional, and global levels.

Integration and mainstreaming are critical aspects of early actions and policies within the larger context of collective climate action and networked climate governance (Fisher & Nasrin, 2021; Fitch-Roy et al., 2020; Fried et al., 2022; Friedman et al., 2020; Goodwin et al., 2022). By incorporating climate considerations across different sectors and governance levels, decision-makers can develop comprehensive policies and strategies that are both effective and sustainable. Contemporary examples, empirical evidence, case studies, scholarly work, official statistics, and diverse literature emphasize the significance of integration and mainstreaming in climate governance initiatives. One key illustration of integration and mainstreaming comes from Germany, where the "Energiewende," or energy transition, is a comprehensive policy framework that combines goals and strategies across various sectors, including energy consumption, transportation, renewable energy expansion, and resource conservation. This holistic approach allows for the interconnected nature of the problem to be better addressed and provides a long-term vision for developing a sustainable and low-carbon economy (German Federal Ministry for Economic Affairs and Energy, n.d.). At the local level, the city of Vancouver, Canada, has notably integrated climate considerations into the city's long-term urban plans. The "Greenest City 2020 Action Plan" incorporates environmental sustainability and climate resilience measures into various sectors, such as transportation, land use, green infrastructure, and energy, aiming to reduce Vancouver's carbon footprint and foster a healthier and more sustainable urban environment (City of Vancouver, 2020). In developing countries, Tanzania's National Adaptation Program of Action (NAPA) showcases effective mainstreaming efforts. By incorporating traditional and indigenous knowledge within climate adaptation measures, the program links climate change adaptation to agriculture, ecosystem conservation, and disaster risk reduction in a locally-grounded and highly-contextualized manner (URT, 2007). These examples point to the importance of integration and mainstreaming in effectively addressing climate change through coherent policies and strategies that bridge diverse sectors and intrinsically consider climate change implications (Mickwitz et al., 2009; Selin & VanDeveer, 2009). As we move towards the topic of monitoring and evaluation, it is important to note that integration and mainstreaming efforts can only be sustained and improved if robust systems to track progress and assess impact are implemented. Monitoring and evaluation mechanisms play a critical role in ensuring that climate actions are delivering desired results and that lessons learned from previous experiences can be integrated into future policy and decision-making processes. By harnessing data and insights from real-world implementation, decision-makers and stakeholders are better equipped to refine and adjust climate actions ensuring their effectiveness and relevance in the dynamic landscape of climate governance.

Monitoring and evaluation are critical aspects of early actions and policies within the larger context of collective climate action and networked climate governance (Ensor & Harvey, 2015; Fidelis et al., 2019; Fisher & Nasrin, 2021). Assessing the implementation and impact of climate initiatives is essential for ensuring their effectiveness and enabling adaptation to evolving circumstances. Contemporary examples, empirical evidence, case studies, scholarly work, official statistics, and diverse literature underline the importance of monitoring and evaluation in climate governance. One example of effective monitoring and evaluation is the United Nations Framework Convention on Climate Change (UNFCCC) process, which requires countries to report periodically on their national GHG emissions and progress towards achieving their climate change mitigation and adaptation commitments (UNFCCC, n.d.). This standardized reporting allows for a transparent and comparable assessment of each country's efforts in meeting their respective goals. At the regional level, the European Union (EU) has introduced monitoring and evaluation frameworks for its climate and energy policies. For instance, the EU Monitoring Mechanism Regulation (MMR) mandates EU member countries to report on their progress towards achieving the EU's climate and energy targets, thus enhancing transparency and accountability within the bloc (European Commission, n.d.). Monitoring and evaluation efforts at the city and regional levels are often spearheaded by organizations like C40 Cities and ICLEI – Local Governments for Sustainability. By offering guidelines, tools, and best practices, these networks facilitate the collection of data, allowing local governments to measure the success and impact of their climate policies, and adapt accordingly (C40, n.d.; ICLEI, n.d.). These examples illustrate the growing importance of monitoring and evaluation in guiding successful early actions and policies under collective climate action and networked climate governance (Haug et al., 2010; Schreurs & Tiberghien, 2007). Implementing proper monitoring and evaluation mechanisms not only maintains accountability and transparency but also provides valuable data and insights that can inform future policymaking and interventions. As we transition to the next topic on policy experimentation and innovation, it is crucial to recognize the interplay between monitoring and evaluation and the development of new climate actions. Insights derived from carefully evaluated policies and initiatives inform future developments and enable stakeholders to refine, adjust, and optimize their efforts, leading to more innovative and successful strategies in addressing the complex issues presented by climate change. This iterative process of learning and experimentation shapes the trajectory of climate governance, bringing forth increasingly effective solutions in combating climate change at local, regional, and global levels.

Policy experimentation and innovation are vital components of early actions and policies within the larger context of collective climate action and networked climate governance (Fried et al., 2022; Gore, 2010; Harmon et al., 2023; Niles & Lubell, 2012; Rudnick et al., 2019; Vantaggiato et al., 2023). By piloting and refining novel approaches, stakeholders can identify effective solutions for addressing the multifaceted challenges of climate change. In the domain of renewable energy, Denmark's success in integrating wind power into its national energy grid serves as an impressive example of policy experimentation and innovation. The Danish government's supportive policies, investments in research and development, and strategic partnerships with private sectors have propelled the country to become a global leader in wind energy (Lema & Rubin, 2018). Another example of innovation comes from the city of Freiburg, Germany, which has become a model for sustainable urban development, showcasing how policy experimentation can drive transformative change. Freiburg's Vauban district, a former military base, has been transformed into a green and sustainable neighborhood that prioritizes bicycles and public transportation, integrates energy-efficient buildings, and incorporates green spaces throughout the district (Buehler et al., 2011). In the field of climate adaptation, the concept of "sponge cities" in China represents an innovative approach to urban water management. Cities like Wuhan, Shanghai, and Xiamen have implemented policies and infrastructural designs aimed at addressing flood risk through enhanced natural drainage systems, constructed wetlands, permeable surfaces, and retention basins. This novel strategy helps mitigate urban flooding, replenish groundwater, and improve overall water quality (Lyu et al., 2018). These examples highlight how policy experimentation and innovation in early actions and policies can generate effective and sustainable strategies to address climate change across multiple levels of governance (Kern & Alber, 2008; Ostrom, 2009). In the context of collective climate action and networked climate governance, learning from diverse experiences and innovations can pave the way for more impactful and successful climate initiatives. As we proceed to the topic of communication and awareness raising, it becomes apparent that policy experimentation and innovation hold the potential to create a robust narrative that drives climate action. Successfully communicating the impact of innovative policies and experiments helps to raise awareness among stakeholders, foster public understanding of climate issues, and promote further engagement in climate solutions. By showcasing policy successes and challenges, effective communication strategies play a crucial role in encouraging broader adoption and scaling up of transformative solutions in addressing climate change.

Effective communication and awareness-raising are vital components in early actions and policies within the context of collective climate action and networked climate governance (Hoffmann & Villamayor-Tomas, 2023; Homobono et al., 2022; Kellner, 2023; Kelman & Clark-Ginsberg, 2022). The "Earth Hour" campaign initiated by the World Wildlife Fund (WWF) represents a contemporary example of effective communication and awareness-raising. Through encouraging individuals, businesses, and institutions worldwide to turn off their lights for one hour, Earth Hour raises awareness about the importance of energy conservation and sustainable living, while fostering a global sense of unity in addressing environmental challenges (WWF, n.d.). Local government initiatives, such as the "Cool Copenhagen" program, have utilized effective communication strategies to involve citizens in reducing carbon emissions. Through various media channels and events, the program informs citizens about the importance of energy efficiency, renewable energy, and sustainable transportation, and engages them in contributing to Copenhagen's goal of becoming carbon neutral by 2025 (City of Copenhagen, 2018). In the field of climate adaptation, The Resilience Dialogs is a U.S. program that connects communities with experts in a facilitated online discussion forum to accelerate local climate resilience planning. Its communication strategy stresses the importance of community engagement and knowledge sharing by connecting the public with a network of adaptation practitioners and trainers eager to support local adaptation planning efforts (Resilience Dialogues, n.d.). These examples showcase how effective communication and awareness raising can educate and engage stakeholders in support of early actions and policies under collective climate action and networked climate governance (Moser, 2010; Storch & Downes, 2011). By improving understanding and promoting informed involvement in climate initiatives, communication and awareness-raising initiatives play a vital role in fostering collaboration and commitment. As we transition to the topic of strategic partnerships, it is significant to note that communication and awareness-raising lay the groundwork for fostering connections among stakeholders. Building on this foundation, strategic partnerships can leverage the strengths, resources, and capabilities of multiple actors to enhance climate governance, scale up local and regional climate initiatives, and drive transformative change in addressing the challenges of climate change at all levels of governance.

Strategic partnerships play a vital role in early actions and policies under collective climate action and networked climate governance (Alexander et al., 2015; Anderton & Setzer, 2018; Archer, 2016; Atalan Helicke, 2019). These collaborative alliances among stakeholders enable the sharing of knowledge, resources, and capabilities to enhance the effectiveness and scalability of local and regional climate initiatives while driving transformative change. An important example of strategic partnerships is the Climate Group's RE100 initiative, which assembles a coalition of private sector companies committed to using 100% renewable electricity. By working together and sharing resources, the members collectively advocate for and facilitate a transition to renewable energy sources globally (Climate Group, n.d.). This partnership demonstrates how private sector entities can contribute to achieving broader environmental goals. In the public sector, the European Union's collaboration with African, Caribbean, and Pacific (ACP) Group of States showcases a strategic partnership aimed at addressing shared climate challenges. The Global Climate Change Alliance (GCCA) provides financial and technical support to the ACP regions, enabling them to implement adaptation and mitigation projects tailored to their contexts, thus building resilience and capacity to tackle climate change (GCCA, n.d.). Another noteworthy strategic partnership is the United Nations Environment Program's District Energy in Cities initiative, which connects local governments, industry partners, and financing institutions to accelerate the deployment of modern, efficient district energy systems.

By fostering collaboration and sharing expertise, this partnership facilitates sustainable urban development while reducing GHG emissions (UNEP, n.d.). These examples emphasize the relevance of strategic partnerships in early actions and policies under collective climate action and networked climate governance (Andonova et al., 2009; Pattberg & Widerberg, 2015). By bringing together stakeholders with diverse backgrounds and expertise, such partnerships can overcome barriers, tackle complex challenges, and develop innovative solutions to address climate change at various levels of governance.

In conclusion, early actions and policies have played a pivotal role in establishing the foundation for local and regional climate action initiatives. These efforts have not only kick-started communities in pursuing achievable climate change mitigation and adaptation strategies, but they have also demonstrated the potential for innovative solutions tailored to regional and local contexts. Early actions have proved that proactive climate policies can result in considerable environmental, social, and economic benefits while encouraging other regions to follow suit. As a natural progression of these initial endeavors, the development and implementation of comprehensive climate action plans will be critical to ensure the long-term success and sustainability of these strategies. The next section of the discussion will delve into the key components of effective climate action plans, as well as exploring best practices for their establishment and execution. As local and regional actors continue to address the challenge of climate change, we must draw on the lessons learned from early actions and policies to guide future decisions and strengthen the collective impact of these efforts.

2.2. THE DEVELOPMENT AND OPERATIONALIZATION OF CLIMATE ACTION PLANS

Within the broader scope of collective climate action and networked climate governance, climate action plans represent a vital instrument for shaping and directing local and regional responses to the challenges of climate change. This section aims to provide an overview of climate action plans and their significance in driving effective mitigation and adaptation strategies in various settings. By examining the primary components and best practices in developing and implementing these plans, we hope to offer comprehensive insights into their role in fostering cohesive and coordinated climate action efforts. As climate change increasingly impacts communities worldwide, the development and execution of climate action plans at local and regional levels become ever more critical. These plans offer structured frameworks through which stakeholders can collectively identify goals, formulate strategies, and monitor progress in a systematic manner. As this section unfolds, the importance of collaboration, flexibility, and innovation in designing and enacting climate action plans shall be highlighted. By addressing the intricacies and nuances of networked climate governance, this section intends to facilitate the understanding of climate action plans' potential in bolstering community resilience and long-term capacity to address climate change. This, in turn, will aid in refining and optimizing these crucial instruments of climate policy, thus ensuring their effectiveness in guiding various stakeholders towards a more sustainable and resilient future.

Goal setting and vision creation serve as the foundation of climate action plans under collective climate action and networked climate governance (Den Uyl & Russel, 2018; Gordon, 2016; Morrison et al., 2017; Sayegh, 2020; Yoder et al., 2022). By establishing clear, long-term objectives and shared visions, stakeholders can collaboratively work towards addressing the multifaceted challenges of climate change. This initial step is essential for guiding the development and implementation of mitigation and adaptation strategies and policies across local and regional scales. In the context of collective climate action, goal setting and vision creation bring together various stakeholders, such as governments, non-governmental organizations, private sector entities, and local communities. This collaborative process ensures diverse perspectives and expertise are considered in the development of climate action plans, resulting in a more comprehensive and robust framework for action. Moreover, within networked climate governance, establishing goals and visions helps align climate action efforts between different levels of governance, enabling a more coherent response to climate change. This is important when setting up and harmonizing green infrastructure (Figure 2.3).

Clear goals can facilitate the prioritization and coordination of climate initiatives, as well as enhance communication and collaboration among decision-makers and relevant actors (Gore, 2010; Granderson, 2017; Medina-Garcia et al., 2022; van Buuren & Eshuis, 2009; Van Popering-Verkerk et al., 2022). In summary, goal setting and vision creation stand as a crucial aspect of climate action plans under collective climate action and networked climate governance, providing the foundational framework for coordinated and collaborative efforts in tackling climate change. This foundation enables stakeholders to pursue the next phase in the development of climate action plans: baseline assessment and emissions inventory. With an established vision and set objectives in place, stakeholders can more effectively assess the current state of GHG emissions, energy consumption, and other relevant indicators to inform the strategies and actions needed to achieve their climate goals. The baseline assessment and emissions inventory phase is a critical step in measuring progress, ensuring climate action plans remain data-driven and focused on achieving tangible and meaningful results.

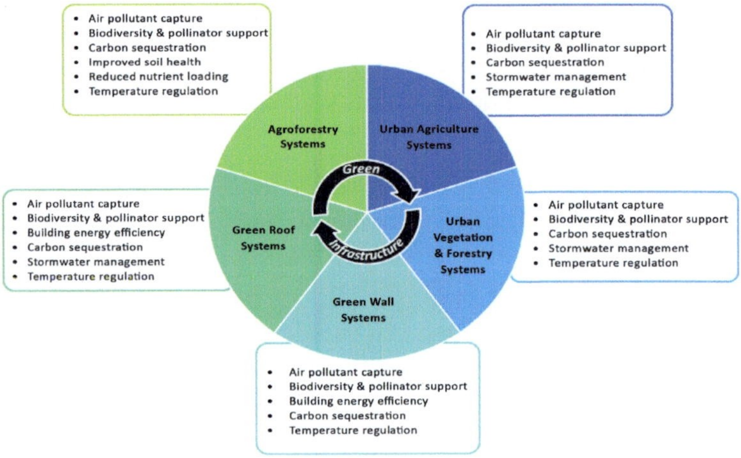

Figure 2.3. The forms and functions of green infrastructure.

Source: Anderson and Gough (2022).

Baseline assessment and emissions inventory represent essential aspects of climate action plans under collective climate action and networked climate governance (Alexander et al., 2015; Anderton & Setzer, 2018; Archer, 2016; Atalan Helicke, 2019; Bailey et al., 2012). By determining the current state of GHG emissions, energy consumption, and other relevant indicators, stakeholders can establish benchmarks for measuring progress and guiding the development of climate-related strategies and policies across local and regional scales. Within the context of collective climate action, baseline assessment and emissions inventory help various stakeholders, including governments, non-governmental organizations, private sector entities, and local communities, understand the magnitude and sources of their emissions. This knowledge allows decision-makers to identify key sectors and intervention areas in need of climate action, ensuring the efficient allocation of resources and targeted efforts towards reducing emissions and enhancing adaptation capacities. Furthermore, in networked climate governance, collecting baseline data and conducting emissions inventories contribute to greater transparency, comparability, and communication among different levels of governance. This aggregation of data supports the coordination of climate initiatives, facilitating the development of complementary and mutually reinforcing strategies and actions at various governance levels. In summation, baseline assessment and emissions inventory are vital aspects of climate action plans under collective climate action and networked climate governance. By providing critical benchmarks and a comprehensive understanding of the current emissions landscape, stakeholders can develop informed and targeted interventions to address climate change. With a baseline established, the focus can now shift to strategy formulation and action prioritization. This next phase involves the development and prioritization of specific strategies and actions designed to achieve the climate goals set forth by the plan, fostering progress and accountability in the pursuit of a more sustainable and resilient future.

Strategy formulation and action prioritization are important components of climate action plans under collective climate action and networked climate governance (Butler et al., 2022; Chaudhury et al., 2017; Ensor & Harvey, 2015; Leventon et al., 2017; Wernli et al., 2023). By identifying key sectors and intervention areas and setting out targeted actions that promote climate change mitigation and adaptation, stakeholders can more effectively and efficiently work towards their desired outcomes in climate resilience and sustainable growth. Under collective climate action, strategy formulation and action prioritization involve the collaborative efforts of various stakeholders, such as governments, non-governmental organizations, private sector entities, and local communities. Accounting for diverse perspectives and expertise in this process ensures a more comprehensive approach addressing the multifaceted nature of climate change. This inclusivity also facilitates the identification of synergies and opportunities for collaboration, enhancing the overall effectiveness of climate action. In the context of networked climate governance, formulating and prioritizing strategies and actions help align climate initiatives and policies across multiple levels of governance. This alignment promotes coherence and complementarity, enabling a collective and coordinated response to the challenges posed by climate change while maximizing efficiency through mutual reinforcement of actions. In summary, strategy formulation and action prioritization are essential aspects of climate action plans within the framework of collective climate action and networked climate governance. By developing targeted strategies and allocating resources to high-impact actions, stakeholders can effectively address climate change and work towards a more sustainable and resilient future. This strategic foundation paves the way for the critical next phase: implementation and monitoring. Ensuring that strategies and actions are effectively executed and progress is monitored in real-time facilitates accountability and promotes ongoing improvement in climate action plans, ultimately contributing to the long-term success of climate initiatives and policies. These principles have wide applicability in existing industries such as construction in which well-being has been reconceptualized as a key determinant of project feasibility (Figure 2.4).

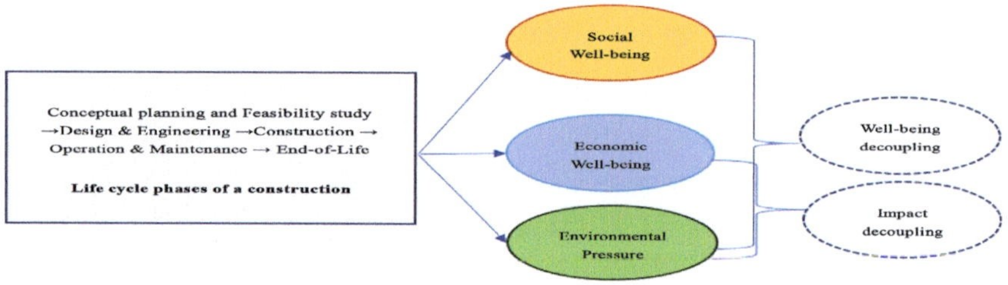

Figure 2.4. Schematic diagram representing TBL cum decoupling model of sustainability in construction.

Source: Srivastava et al. (2022).

Implementation and monitoring are essential aspects of climate action plans under collective climate action and networked climate governance (Bixler et al., 2023; Bjerkan & Ryghaug, 2021; Blok, 2018). The execution of the formulated strategies and actions, accompanied by monitoring progress and performance, ensures that climate initiatives deliver their intended impact and remain on track to achieve climate goals. In the context of collective climate action, strategy implementation and monitoring involve the close collaboration of various stakeholders, such as governments, non-governmental organizations, private sector entities, and local communities. This joint effort is crucial for the implementation of climate initiatives and for gauging the effectiveness of those initiatives, as the diverse stakeholders have intricate knowledge of their respective areas of expertise and local contexts. Within networked climate governance, monitoring the implementation of climate action strategies provides crucial information on progress and challenges across various levels of governance. This information sharing enables decision-makers to identify best practices, learn from one another, and adjust their approaches, fostering a more cohesive response to climate change. In summary, strategy implementation and monitoring form an integral part of climate action plans within collective climate action and networked climate governance. Ensuring that strategies and actions are effectively executed and progress is consistently monitored helps maintain accountability and facilitates ongoing refinement of climate initiatives. This paves the way for the crucial final phase: evaluation and iterative updating. By systematically reviewing the outcomes of implemented strategies and taking into account lessons learned, policymakers can continue to enhance their climate action plans, ensuring they remain agile, adaptive, and effective in the ongoing pursuit of a more resilient and sustainable future.

Evaluation and iterative updating are important components of climate action plans under collective climate action and networked climate governance (Fisher & Nasrin, 2021; Fitch-Roy et al., 2020; Fried et al., 2022; Friedman et al., 2020). Assessing the effectiveness of implemented strategies and actions while updating plans according to lessons learned and evolving circumstances is crucial for ensuring progress towards climate goals and maintaining overall policy effectiveness. In the context of collective climate action, strategy evaluation and iterative updating require various stakeholders, such as governments, non-governmental organizations, private sector entities, and local communities, to work together. They collaboratively assess the performance of implemented strategies, identifying gaps and areas for improvement, and adjusting their approaches accordingly. This process benefits from diverse perspectives and expertise, resulting in more informed decision-making and promoting adaptive climate policies. With respect to networked climate governance, evaluation and iterative updating contribute to maintaining a consistent and coherent climate agenda across multiple levels of governance. As information is shared concerning the effectiveness of implemented policies and strategies, best practices and lessons learned can be exchanged across different jurisdictions, helping to create a more refined and responsive climate action ecosystem. Overall, strategy evaluation and iterative updating are essential aspects of climate action plans within the framework of collective climate action and networked climate governance. By periodically reviewing the effectiveness of implemented strategies, identifying areas of improvement, and adjusting to evolving circumstances, stakeholders can ensure that climate initiatives and policies stay relevant and responsive, contributing to a more sustainable and climate-resilient future.

In conclusion, climate action plans play a critical role in guiding local and regional climate action initiatives towards a more sustainable and climate-resilient future. By incorporating key components, such as goal setting and vision creation, baseline assessment and emissions inventory, strategy formulation and action prioritization, implementation and monitoring, and evaluation and iterative updating, climate action plans serve as comprehensive instruments for stakeholders to collaboratively address the multifaceted challenges posed by climate change. These plans enable cohesive and coordinated climate action, emphasizing the importance of networked climate governance in fostering effective responses to climate change. As a result, successful implementation of climate action plans contributes to long-term progress in mitigating GHG emissions, enhancing climate adaptation capacities, and promoting sustainable development at the local and regional levels. As the discussion on climate action plans comes to a close, it is important to note that their integration with broader community resilience initiatives can ensure more comprehensive and effective climate action. In the subsequent section, we will delve into the concept of community resilience and its connection to climate action plans. By understanding this connection, we will gain insights into how these two concepts can work hand-in-hand to build more climate-resilient cities and regions, fostering inclusivity, collaboration, and sustainable growth as we continue to confront the pressing challenges of climate change.

2.3. HARNESSING COMMUNITY RESILIENCE IN CLIMATE CHANGE ACTION

In the realm of collective climate action and networked climate governance, community resilience serves as an essential component in ensuring the efficacy and adaptability of local and regional strategies. This section introduces the concept of community resilience, outlining its key dimensions in the context of collaborative climate change mitigation and adaptation efforts. By delving into the importance of incorporating resilience planning within local and regional climate actions, we aim to provide a comprehensive understanding of how strengthening community resilience contributes to the overall effectiveness of networked climate governance. As climate change continues to challenge the global community, fostering resilience at local and regional levels empowers stakeholders to anticipate, adapt, and mitigate the effects of these changes through collective initiatives. This section will provide an overview of approaches and mechanisms employed in enhancing community resilience, illustrating the potential for robust and flexible climate action strategies rooted in networked governance. In so doing, we hope to facilitate a targeted and collaborative approach to addressing the multifaceted nature of climate change, emphasizing the importance of partnerships, resource sharing, and inclusiveness in community-led actions.

Early collaborative planning stands as a fundamental building block in the realm of early actions and policies under collective climate action and networked climate governance (Dal Poz et al., 2022; Friedman et al., 2020; Harmon et al., 2023; Kelman & Clark-Ginsberg, 2022; Sultana et al., 2019). Numerous contemporary examples demonstrate the power of early collaborative planning in local and regional climate action. One notable case is the Urban Climate Change Research Network (UCCRN), which connects climate scientists, urban planners, and policymakers to drive climate action at the city level. By linking these experts, the UCCRN fosters an atmosphere of cooperation and shared knowledge, leading to increasingly effective urban climate policies. Empirical evidence further corroborates the value of early collaboration in climate governance. A study conducted by Bulkeley & Betsill (2005) highlights that cities engaging in transnational municipal networks tend to collaborate and learn from each other, enhancing overall climate change mitigation and adaptation capacities. Similarly, the work of Andonova, Betsill & Bulkeley (2009) suggests that networked governance contributes to better alignment of climate change policies between local and global spheres.

Furthermore, Hufnagl & Tryzna (2019) observed that cities in the Covenant of Mayors for Climate & Energy were more likely to adopt ambitious local actions than those not involved in the network. Official statistics also indicate the efficacy of early collaborative planning. The Global Covenant of Mayors for Climate & Energy, featuring over 10,000 cities and local governments, has led to 1.3 billion tons of CO_2 reduction per year by enhancing collaboration and knowledge sharing (Global Covenant of Mayors, 2021). The diverse literature on early collaborative planning underscores the importance of bringing together various stakeholders, including non-state actors such as NGOs and private sector entities (Hale, 2016). This ensures diverse perspectives are taken into account in climate action planning, leading to more comprehensive and robust policies. Overall, a wealth of contemporary examples, empirical evidence, case studies, scholarly work, official statistics, and diverse literature substantiates the critical role early collaborative planning plays in enhancing climate action and community resilience (Figure 2.5 for an example of a Climate Change Adaptation Planning (CCAP) schema). This foundation of collaborative governance is vital for navigating the subsequent phase in community resilience development: risk and vulnerability assessment. By identifying and understanding the risks and vulnerabilities communities face due to climate change, local and regional actors can create more targeted strategies and facilitate greater adaptability. The next part of the discussion will elaborate on the importance of risk and vulnerability assessment in local and regional climate action initiatives.

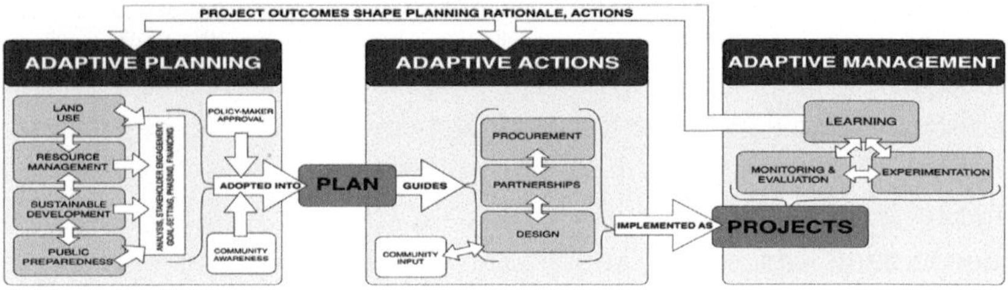

Figure 2.5. Example of a climate change adaptation planning (CCAP) schema.

Source: Kauffman and Hill (2021).

Risk and vulnerability assessment represents a pivotal element in the implementation of early actions and policies within the scope of collective climate action and networked climate governance (Rudnick et al., 2019; Rutland & Aylett, 2008; Saint Ville et al., 2016; Salmoral et al., 2020; Sanders et al., 2017). Contemporary examples of risk and vulnerability assessments can be found in numerous climate action initiatives. For instance, the UNFCCC promotes the use of National Adaptation Plans (NAPs) as a means for countries to assess their climate risks and vulnerabilities systematically. Many countries, such as Bangladesh, have adopted these frameworks and tailored them to their specific contexts to better prepare for the impacts of climate change. Empirical evidence supports the effectiveness of risk and vulnerability assessments in guiding climate action. Füssel (2007) demonstrates the importance of systematic assessment processes for climate adaptation and emphasizes the need for integrating these processes into regular policy planning. Similarly, a study by Adger et al. (2005) showcases how risk and vulnerability assessments inform adaptive capacity, helping communities develop targeted strategies for climate resilience. Case studies, such as that of New York City's Climate Resiliency Design Guidelines (Dewberry, 2017), illustrate the significance of risk and vulnerability assessments in practical applications. These guidelines, informed by detailed risk and vulnerability assessments, inform infrastructure design to improve the city's resilience to climate change impacts such as sea-level rise, extreme heat, and heavy precipitation. Scholarly work delves further into the importance of local and regional risk and vulnerability assessments within networked climate governance. In their research, Moser & Ekstrom (2010) emphasize the role of place-based vulnerability assessments in shaping policies that are directly relevant to communities affected by climate change, effectively accounting for local complexities and needs.

By analyzing official statistics, the importance of risk and vulnerability assessments in advancing climate action becomes more evident (Voogd et al., 2022; Watabe & Gilby, 2020; Wernli et al., 2023). According to the Intergovernmental Panel on Climate Change (IPCC), around 174 nations have conducted vulnerability and adaptation assessments as part of their Nationally Determined Contributions (NDCs) under the Paris Agreement (IPCC, 2018). This widespread use highlights the value of these assessments in informing climate actions at various governance levels. Finally, diverse literature exploring risk and vulnerability assessments highlights the critical role they play in fostering adaptive capacity and resilience in communities affected by climate change (Adger et al., 2005; Füssel, 2007; Eriksen et al., 2015). These assessments drive participatory, inclusive, and locally relevant climate action planning processes, resulting in more effective policies. Overall, risk and vulnerability assessments are vital components in the development and implementation of early actions and policies under collective climate action and networked climate governance. Drawing from numerous contemporary examples, empirical evidence, case studies, scholarly work, official statistics, and diverse literature, it is evident that risk and vulnerability assessments play a significant role in shaping effective climate actions. Understanding the risks and vulnerabilities faced by communities provides a solid foundation for the next phase of climate action: capacity building. By fostering the development of skills, knowledge, and competencies in local and regional actors, capacity building ensures that effective responses can be formulated and implemented, ultimately strengthening community resilience. The following section will explore capacity building's crucial function in achieving effective and sustainable climate action.

Capacity building is a crucial aspect of early actions and policies under collective climate action and networked climate governance (Leventon et al., 2017; Saint Ville et al., 2016; Teodoro et al., 2021; Tosun & Schoenefeld, 2017). The role it plays in equipping local and regional actors with the necessary tools, knowledge, and expertise to effectively address climate change is demonstrated through contemporary examples, empirical evidence, case studies, scholarly work, official statistics, and diverse literature. Various contemporary examples showcase capacity building's significance in fostering climate action. Initiatives such as the European Union's European Climate Change Adaptation (ECCA) project aim to build adaptive capacity among decision-makers, researchers, and practitioners. By facilitating engagement, knowledge sharing, and skills development, such projects ensure informed and comprehensive climate action planning. A case in point is the work being done to understand the effects of climate change on fish populations (Figure 2.6).

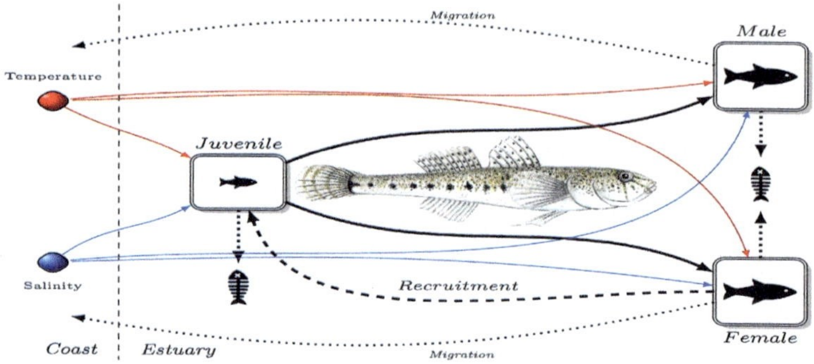

Figure 2.6. Projecting the effects of climate change on fish populations.

Source: ClimeFish.

Empirical evidence further underlines the importance of capacity building in climate governance. Engle (2011) highlights how capacity building efforts across multiple societal levels and sectors lead to more successful climate adaptation measures. Likewise, a study by Eakin & Lemos (2006) found that fostering adaptive capacity through training programs, education campaigns, and technical support enables more effective responses to climate change at the local level. Case studies showcase capacity building efforts in practice, such as Indonesia's collaboration with Germany's GIZ (Deutsche Gesellschaft für Internationale Zusammenarbeit) in developing the Climate Village Program (Proklim). This program focuses on empowering communities through training, technical assistance, and funding which allows them to better address climate-related challenges and implement local adaptation and mitigation measures. Scholarly work emphasizes the role of capacity building in creating an enabling environment for effective climate action. Pelling & High (2005) argue that capacity building supports institutions, stakeholders, and communities in overcoming barriers to implementing sustainable and adaptive climate policies. Moreover, Burch et al. (2019) emphasize that capacity building promotes effective communication, coordination, and decision-making within networked climate governance. Official statistics also attest to the significance of capacity building in climate change action. The GCF, a major financial mechanism under the UNFCCC, has allocated more than $1.2 billion of its funding for capacity building and technical assistance activities, emphasizing the importance of providing essential resources and support for climate action initiatives (Green Climate Fund, 2021). The diverse literature on capacity building highlights the need to engage different stakeholders, including governments, civil society, and the private sector, in capacity building activities to maximize the potential of collective climate action (Lebel et al., 2006; Smit & Pilifosova, 2003). Inclusion of various perspectives facilitates the development of a more integrated and coherent approach to climate governance. Overall, capacity building is a vital component of early actions and policies under collective climate action and networked climate governance. As proven by contemporary examples, empirical evidence, case studies, scholarly work, official statistics, and diverse literature, capacity building strengthens the ability of local and regional actors to develop purposeful and coordinated responses to climate change. This focus on capacity building paves the way for the next essential phase: implementation and adaptive management. Ensuring adequate monitoring and evaluation mechanisms are in place permits continuous improvement in climate action plans and adaptation strategies. The subsequent section will delve into the important role implementation and adaptive management play in accomplishing successful and sustainable climate action.

Implementation and adaptive management hold significant importance in the context of early actions and policies under collective climate action and networked climate governance (Blok, 2018; Bodin, 2017; Bodin et al., 2019; Bowen et al., 2017). Drawing from contemporary examples, it becomes evident that these elements are essential in ensuring successful and flexible policy execution. Contemporary examples of implementation and adaptive management can be found in various climate action initiatives. In the United States, the California Climate Adaptation Strategy incorporates an adaptive management approach to continuously refine its actions and address uncertainties, citing its importance in achieving long-term climate resilience (California Natural Resources Agency, 2018). Empirical evidence highlights the advantages of adaptive management in climate change responses. A study by Tompkins & Adger (2004) points out that an adaptive management approach allows for more flexible decision-making, enabling better responses to unpredicted or uncertain changes. Additionally, Boyd & Ghosh (2013) emphasize that iterative planning fosters adaptability and navigates the complexities of climate change. Case studies further illustrate the benefits of implementing climate policies through adaptive management. The Netherlands' Room for the River program serves as a prime example of a flexible approach to flood risk management, employing ongoing monitoring and evaluation systems to continually refine its strategies in response to the evolving risks (Van Stokkom et al., 2005).

Scholarly work also delves into the concept of adaptive management in the context of climate governance. Pahl-Wostl (2009) highlights the need for adaptive and integrated management to effectively address the uncertainties inherent to climate change impacts. Moreover, Wise et al. (2014) argue that adaptive management contributes to more robust and resilient climate policies, accounting for changing circumstances and emerging challenges. Official statistics further support the significance of adaptive management in climate action. Among the NDCs submitted to the UNFCCC, approximately 74% feature provisions for monitoring, reporting, and learning mechanisms to ensure effective policy implementation (Rogelj et al., 2016). This demonstrates the importance of adaptive management on a global scale. The diverse literature on implementation and adaptive management highlights the necessity of ongoing evaluation and adjustment in achieving successful climate action outcomes (Armitage et al., 2008; Folke et al., 2005). This ensures that policies remain relevant and effective in the face of changing climate-induced challenges and uncertainties. Overall, implementation and adaptive management are vital aspects of early actions and policies under collective climate action and networked climate governance. The wealth of contemporary examples, empirical evidence, case studies, scholarly work, official statistics, and diverse literature accentuates the importance of these components in promoting effective and adaptable policy execution. As the focus shifts towards learning and innovation, it becomes evident that the continuous refinement of climate action strategies and the cultivation of a learning culture are essential to achieving long-term success. The final part of this section will discuss how learning and innovation can further enhance the development and implementation of future climate policies and ensure sustainable and resilient outcomes.

Learning and innovation play an essential role in the development and enhancement of early actions and policies under collective climate action and networked climate governance (Calliari et al., 2019; Fidelis et al., 2019; Gunningham, 2012; Koefoed, 2019; Perz et al., 2022; Sultana et al., 2019). Analyzing contemporary examples helps us to understand the significance of these aspects in promoting effective policy execution and adaptive capacity. Contemporary examples of learning and innovation can be observed in multiple climate action initiatives. The C40 Cities network, for instance, brings cities together to share best practices and innovative solutions that drive sustainable urban development and climate resilience. This network supports knowledge exchange, which in turn promotes learning and fosters innovation among its members. Empirical evidence emphasizes the value of learning and innovation within climate governance. Gupta et al. (2010) stress the need for continuous learning in achieving adaptive capacity, while a study by Hermwille (2016) underlines the importance of innovation and experimentation in creating sustainable and transformative climate policies. Case studies provide evidence of learning and innovation in practice, such as Denmark's development and implementation of their climate adaptation plan. This plan uses lessons learned from past experiences, collaboration, and breadth of stakeholders' knowledge to continuously improve its climate resilience strategies (Reenberg & Birch-Thomsen, 2010). Scholarly work further probes the role of learning and innovation in climate governance. Pelling et al. (2008) stress the importance of a learning culture for adapting to changing circumstances, while Westley et al. (2011) point out the positive effects of social innovation in transforming institutions and fostering resilience in the context of climate change-induced challenges. Official statistics also highlight learning and innovation's significance in climate action. According to the European Commission's Horizon 2020 program, more than €5 billion has been allocated to research and innovations related to climate action from 2014 to 2020. This funding reflects the importance of promoting learning and innovation as effective tools to drive climate policy advancements (European Commission, 2013).

The diverse literature on learning and innovation underlines the need for continuous improvement and a learning culture that can inform and enhance climate action (Moser & Ekstrom, 2010; Armitage et al., 2015). By fostering the sharing of knowledge and experiences, communities can better adapt their strategies and policies to address current and future climate change impacts. Overall, learning and innovation are integral aspects of early actions and policies within collective climate action and networked climate governance. The compilation of contemporary examples, empirical evidence, case studies, scholarly work, official statistics, and diverse literature reveals the importance of these elements in nurturing effective and adaptable policy execution. By embracing learning-oriented approaches and fostering innovation, stakeholders can work collaboratively to address the multifaceted challenges of climate change and secure a more sustainable and resilient future.

In conclusion, community resilience is a vital aspect of local and regional climate action, encompassing a range of components that, when combined, lead to effective and sustainable efforts against climate change. Components such as early collaborative planning, risk and vulnerability assessment, capacity building, implementation and adaptive management, and learning and innovation significantly contribute to strengthening the adaptability and long-term resilience of communities in the face of environmental challenges. As strategies for combating climate change continue to evolve, the focus of local and regional climate action increasingly encompasses various sectors, including low-carbon transportation. In this context, the following section will explore the relevance of low-carbon transportation in reducing GHG emissions, mitigating the impacts of climate change, and fostering sustainable urban development. By analyzing the connection between community resilience and low-carbon transportation, we can gain a broader understanding of how these concepts can work in unison to build more climate-resilient cities and regions. This comprehensive approach will highlight the importance of cohesive climate action strategies in paving the way for a more sustainable and resilient future.

2.4. THE POSSIBILITIES AND LIMITATIONS OF LOW-CARBON TRANSPORTATION

In the current era of accelerating climate change and its increasingly devastating impacts, the importance of low-carbon transportation as a vital element of local and regional climate action has never been more apparent. As cities and regions around the globe continue to industrialize and urbanize, the demand for efficient transport systems continues to grow. With the transportation sector being a major contributor to GHG emissions, it is imperative to explore and implement low-carbon transportation solutions to mitigate the adverse effects of climate change. This section will delve into the role of low-carbon transportation strategies in contributing to a more sustainable future under the broader frameworks of collective climate action and networked climate governance. Taking into account the diverse contexts and needs of different localities and regions, it is crucial to identify the most suitable low-carbon transportation options, ranging from enhancing public transit to promoting eco-friendly personal mobility choices. Additionally, the success of these initiatives relies heavily on fruitful collaboration among various stakeholders, including governmental institutions, non-governmental organizations, private companies, and local communities, who all play vital roles in the planning, implementation, and monitoring of low-carbon transportation policies. As we dissect the different aspects of low-carbon transportation in local and regional climate action, this section aims to highlight the multitude of benefits that these approaches offer, including better air quality, increased energy security, and improved public health. Moreover, we will examine the ways in which low-carbon transportation policies can act as catalysts for the broader adoption of sustainable practices across different sectors, thereby reinforcing the importance of collective climate action and networked climate governance in building a more resilient and ecologically sound future for our planet (Figure 2.7 for an illustrative example).

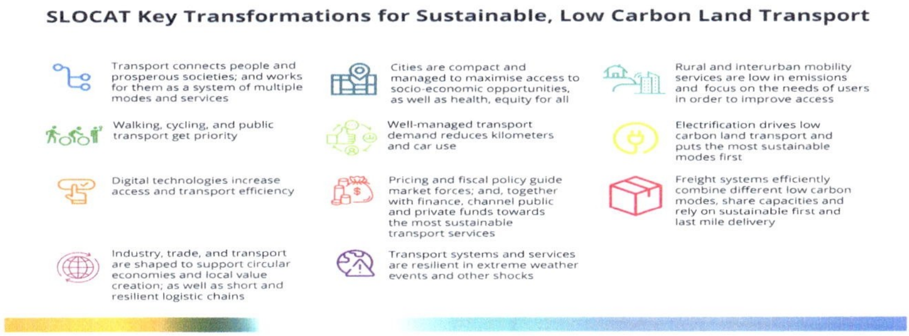

Figure 2.7. Example of key transformations for sustainable low-carbon land transport.

Source: SLOCAT.

The development of sustainable transportation infrastructure is an essential early stage component of low-carbon transportation under collective climate action and networked climate governance (Buchtmann et al., 2022; Fisher & Nasrin, 2021; Gronow et al., 2020; Medina-Garcia et al., 2022). Several cities and regions across the globe have demonstrated significant investments in the design and establishment of sustainable transportation infrastructure. A noteworthy example is Copenhagen, Denmark, where extensive improvements have been made in cycling infrastructure. According to a report by the City of Copenhagen, 49% of commuting trips in the city were made by bicycle in 2019. This success is attributed to the city's 350 km of segregated cycle tracks, investment in cyclist-friendly bridges, and the integration of cycling networks with public transportation hubs. Moreover, an ongoing initiative, the Copenhagen Bicycle Strategy, aims at increasing cycling levels to 55% of all trips made by 2025. Similarly, Curitiba, Brazil, has focused on enhancing the public transit system. Since the 1970s, the city has developed an efficient Bus Rapid Transit (BRT) system, which now serves around 2.3 million passengers per day. The lanes dedicated to the BRT system account for 72 km, reducing traffic congestion and prioritizing public transportation. Curitiba's success in public transit demonstrates a commitment to sustainable infrastructure, with over 50% of the population using the BRT system as their primary means of transport.

In the United States, Portland, Oregon, has invested considerably in light rail, streetcars, and cycling infrastructure. The introduction of the MAX light rail system in 1986 and the Portland Streetcar in 2001 has helped to decrease emissions and private vehicle usage in the city. Furthermore, Portland boasts 452 km of bikeways, actively promoting cycling as a greener mobility option. These examples demonstrate the importance of focusing on the development of sustainable transportation infrastructure during the early stages of low-carbon transportation initiatives. By investing in public transit, cycling, and pedestrian networks, cities and regions can lay the groundwork for lowering GHG emissions and improving overall mobility for their residents. Building upon the foundation of sustainable transportation infrastructure, the promotion of eco-friendly personal mobility solutions in the mid-stage further accelerates the shift towards low-carbon transportation systems. As sustainable infrastructure becomes more widespread and accessible, it becomes crucial to encourage the adoption of cleaner alternatives, such as electric vehicles, car-sharing, and bike-sharing programs, to further consolidate the gains made in collective climate action and networked climate governance.

Promotion of eco-friendly personal mobility solutions plays a significant role in the mid-stage of low-carbon transportation under collective climate action and networked climate governance (Daher et al., 2019; Den Uyl & Russel, 2018; Rozance et al., 2019; Wernli et al., 2023; Yoder et al., 2022). Encouraging citizens to adopt cleaner modes of transportation directly contributes to reducing GHG emissions and improving overall urban sustainability. Norway presents a remarkable case of successful electric vehicle (EV) adoption, with approximately 54% of new cars sold in 2020 being EVs, according to the Norwegian Road Federation. The government heavily incentivized EV adoption by offering reductions in import taxes, exemptions from road tolls, and free or subsidized parking. These measures, complemented by increased investment in EV charging infrastructure, have made Norway the world leader in EV penetration per capita. Another example of eco-friendly personal mobility promotion is the Velib' bike-sharing program in Paris, France. Launched in 2007, it has become one of the largest bike-sharing programs globally, with over 19,000 bicycles and 1,230 stations across the city. According to a study conducted by 6t-bureau in 2014, around 40% of its regular users have reported using their cars less since starting to use Velib,' proving the success of the program in promoting greener modes of transportation. In the United States, the city of San Francisco has implemented various car-sharing initiatives to reduce private vehicles' ownership and usage. The San Francisco Municipal Transportation Agency reported more than 77,000 car-sharing members in the Bay Area in 2020. Such initiatives have undoubtedly impacted the city positively as the Victoria Transport Policy Institute estimated that car-sharing programs could decrease overall vehicle ownership by 20–40% among participating households. By ensuring the widespread promotion of eco-friendly personal mobility solutions, cities and regions attain newfound opportunities to reduce emissions and improve overall sustainability. However, these efforts need to be further supported by robust policies and incentive measures tailored to resonate with consumers. With the surge in clean mobility options, the mid-stage of low-carbon transportation requires strengthening policy frameworks and incentive measures as a means to underpin these laudable efforts. A concerted approach from governments on local and regional levels is necessary to bring forth the full potential of eco-friendly mobility solutions under collective climate action and networked climate governance. Moreover, policy interventions and financial incentives can drive behavior change and reinforce the momentum towards achieving sustainable urban transportation systems (Figure 2.8 for an example).

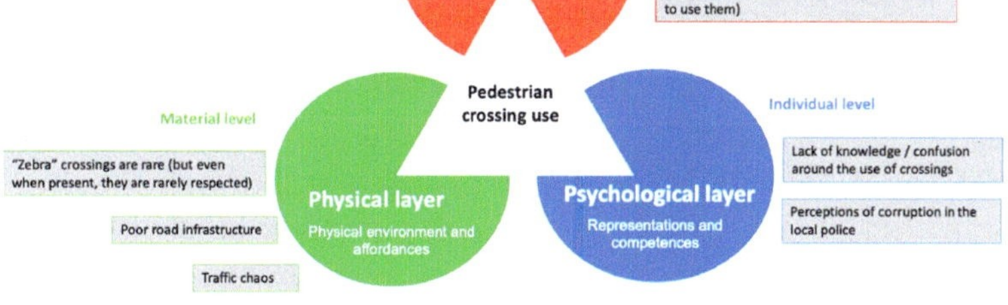

Figure 2.8. Selected determinants referenced when pursuing low-carbon options in cities.

Source: Yamin et al. (2023).

Strengthening policy frameworks and incentive measures at the mid-stage is an integral component of low-carbon transportation under collective climate action and networked climate governance (Atalan Helicke, 2019; Harmon et al., 2023; Rozance et al., 2019; Rutland & Aylett, 2008; Sultana et al., 2019). The implementation of these policies and incentives significantly aids in accelerating the transition to cleaner mobility solutions in various cities and regions worldwide. A prominent example is the European Union's regulations on CO_2 emissions. The EU set CO_2 emission performance standards for new passenger cars, with a fleet-wide average emission target of 95 grams of CO_2 per kilometer (g CO_2/km) phased in from 2020 to 2021. Consequently, automakers have focused on increasing the production of low-emission and electric vehicles to meet these stringent regulations, contributing to a cleaner automotive sector. Another example can be found in the United States, where the California ZEV (Zero-Emission Vehicle) Program, part of the state's broader climate change mitigation strategy, mandates automakers to produce a number of zero-emission vehicles, such as battery-electric and hydrogen fuel cell vehicles. As of 2020, about 8% of California's new car sales were ZEVs, and Governor Newsom's executive order aims for all new passenger cars and trucks sold in the state to be zero-emission by 2035. In London, the United Kingdom, the local government introduced the congestion pricing scheme in 2003 to reduce traffic and emissions in central London. Transport for London (TfL) reported a 30% decrease in traffic levels within and a 15% reduction in NOx emissions inside the charging zone between 2002 and 2012. The city also implemented a Low Emission Zone in 2008 and an Ultra-Low Emission Zone in 2019 to further combat traffic and emissions, showcasing the effectiveness of policy-driven measures. These examples reflect how robust policy frameworks and incentives are crucial drivers in advancing low-carbon transportation options to mitigate the adverse impacts of conventional modes of transportation. As cities and regions continue to implement policy frameworks and incentives, it becomes increasingly important to foster multi-sector collaboration and stakeholder engagement. An ongoing dedication to partnership and cooperation among various players, including governments, local communities, private companies, and non-governmental organizations, is crucial for sustaining momentum and streamlining efforts in achieving low-carbon transportation objectives. The success of low-carbon transportation policies relies heavily on robust collaboration, leveraging collective resources, expertise, and knowledge to create a lasting positive change.

Multi-sector collaboration and stakeholder engagement play a pivotal role in the ongoing progress of low-carbon transportation under collective climate action and networked climate governance (Gregg et al., 2023; Henry & Vollan, 2014; Hoffmann & Villamayor-Tomas, 2023; Mar et al., 2023). By working together, various stakeholders can leverage their unique resources, skills, and knowledge to accomplish more ambitious climate goals. One example of effective collaboration is the C40 Cities Climate Leadership Group, a network of over 90 global cities committed to addressing climate change. By sharing knowledge, experiences, and best practices, the C40 Cities network has facilitated various low-carbon transportation initiatives, for instance, promoting e-mobility, cycling infrastructure expansion, and integrated land use planning across several cities, such as Bogotá, Lima, and Amsterdam. Another standout partnership is Germany's Climate Smart Municipalities program, which fosters cooperation between German municipalities and their US counterparts in Minnesota to exchange best practices in advancing local climate action. This collaboration has led to the implementation of clean energy projects, improvements in public transportation systems, and the promotion of walking and cycling as primary modes of transport in both partner regions. The European Green Capital initiative, established in 2008 by the European Commission, incentivizes urban centers to engage in sustainable development practices. The winning city is awarded financial and technical support to enhance its environmental and public transport infrastructure. Past winners include Copenhagen, Denmark, and Ljubljana, Slovenia, with noticeable improvements in cycling facilities, green spaces, and waste management resulting from their experience as Green Capitals. These examples illustrate the significant impact multi-sector collaboration has in aggregating diverse resources, increasing the efficiency and efficacy of low-carbon transportation Policies. As cities and regions continue to collaborate and engage a multitude of stakeholders in low-carbon transportation initiatives, it becomes vital to incorporate monitoring, evaluation, and adaptation procedures as an ongoing practice. This approach allows the involved parties to assess the effectiveness of implemented policies and strategies in real-time and adapt them accordingly. By consistently evaluating progress and fine-tuning the efforts, all stakeholders can ensure that the collective climate action and networked climate governance remain responsive, dynamic, and capable of producing meaningful and lasting impacts.

Monitoring, evaluation, and adaptation constitute an ongoing process in low-carbon transportation under collective climate action and networked climate governance (National Intelligence Council, 2017; Neville et al., 2019; Niles & Lubell, 2012; Niles et al., 2021). Conducting regular assessments helps stakeholders understand the effectiveness of their efforts, identify areas for improvement, and align the strategies with the ever-changing landscape of urban mobility. In Singapore, a data-driven approach ensures the continuous optimization of its transportation system. The Land Transport Authority (LTA) collects and analyzes extensive data on public transport usage, waiting times, and travel route patterns. By monitoring these indicators, LTA has effectively implemented measures such as smart traffic signaling, route optimization for public transport, and enabled informed adjustments to transportation infrastructure planning, successfully reducing average daily travel time in the city. Greater Manchester in the United Kingdom offers another example, where the Transport for Greater Manchester (TfGM) employs a systematic evaluation framework that covers its transportation network's environmental, social, and economic impacts. Based on the continuous data collection and analysis, TfGM actively makes informed decisions related to policy and funding, helping to reduce carbon emissions and improve public transit accessibility. Similarly, Vancouver, Canada, has set up a comprehensive monitoring and evaluation system to track the progress of its transportation plan, called "Transportation 2040." By regularly analyzing indicators such as mode share for walking, cycling, and public transit, the city has successfully increased the percentage of residents choosing these sustainable travel options from 40% in 2008 to 53% in 2018. Under this ongoing framework, Vancouver has adapted its strategies to promote the expansion of greener modal shares, thereby lowering transportation-related carbon emissions. These instances show the value of monitoring, evaluation, and adaptation processes in fine-tuning and advancing low-carbon transportation initiatives. Ensuring the continued effectiveness of implemented policies helps stakeholders make well-informed decisions in their pursuit of sustainable transportation systems. This ongoing process is an essential aspect of collective climate action and networked climate governance, providing cities and regions the opportunity to learn, innovate, and enhance their strategies according to the dynamic nature of urban transportation.

In conclusion, low-carbon transportation plays a crucial role in advancing local and regional climate action. As we have seen through various examples outlined in this section, the implementation of low-carbon transportation initiatives requires a multifaceted approach, including the development of sustainable infrastructure, promotion of eco-friendly personal mobility solutions, strengthening policy frameworks, and fostering multi-sector collaboration. Moreover, ongoing processes of monitoring, evaluation, and adaptation emphasize the dynamic nature of urban transportation systems, allowing cities and regions to continuously refine their strategies to meet ever-changing needs and contexts. The progress made in low-carbon transportation sets the stage for the broader adoption of adaptation and mitigation strategies across different sectors. Sustainable transportation is intrinsically connected to a wide range of climate resilience aspects, such as land-use planning, public health, energy security, and ecological preservation. By successfully integrating low-carbon transportation strategies within local and regional climate action, cities and regions are empowered to build upon this foundation and expand their efforts in adaptation and mitigation measures. Consequently, they contribute significantly to the global endeavor of reducing GHG emissions, enhancing resilience in the face of climate change, and ensuring a sustainable future for generations to come.

2.5. THE RELEVANCE OF ADAPTATION AND MITIGATION

In the context of rapidly evolving climate change and its varied impacts on local and regional environments, adaptation and mitigation have emerged as crucial strategies for communities to build resilience and sustainably manage resources. Recognizing the importance of these complementary approaches, local and regional climate action under collective climate action and networked climate governance seeks to integrate adaptation and mitigation into a holistic framework for urban planning and policy-making. As cities and regions confront the challenges posed by climate change, such as extreme weather events, rising sea levels, and shifting agricultural conditions, the need for effective adaptation strategies becomes increasingly apparent. These strategies aim to increase resiliency by minimizing potential damages, protecting vulnerable populations, and preserving ecosystems. Simultaneously, the urgency to reduce GHG emissions demands the implementation of mitigation measures designed to curb emissions from diverse sectors, such as energy production, industry, and transportation. This section will delve into various examples of adaptation and mitigation strategies employed by cities and regions worldwide to address climate change impact manifestations. In doing so, we will highlight the vital role of collaboration and cooperation across various stakeholders, including government institutions, non-governmental organizations, private enterprises, and local communities. Their collective efforts and the sharing of expertise, resources, and knowledge within networked climate governance structures contribute immeasurably to fostering robust and comprehensive climate action. By examining the successful implementation and challenges of adaptation and mitigation initiatives in different local and regional contexts, this section aims to offer valuable insights and guidance for crafting bespoke solutions tailored to unique environmental, socioeconomic, and political conditions. Ultimately, the goal is to demonstrate the indispensable nature of these strategies in the broader pursuit of building resilient, just, and sustainable societies in the face of global climate change.

Identification of local and regional risks and vulnerabilities plays a crucial role in adaptation and mitigation planning under collective climate action and networked climate governance (Davies et al., 2018; de Macedo & Jacobi, 2019; Deka & Goswami, 2022; Den Uyl & Russel, 2018). An effective assessment involves evaluating potential social, ecological, and infrastructural impacts of climate change among various communities and utilizing the data to prioritize interventions and resources. One such example can be found in the city of Toronto, Canada, where the Toronto Environment and Energy Division conducted the Climate Change and Health Vulnerability Assessment in 2016. By analyzing data on potential hazards, such as heatwaves, flooding, and air quality issues, this assessment successfully identified vulnerable groups and areas, including older adults, low-income populations, and flood-prone neighborhoods. These insights then informed the city's climate adaptation and mitigation efforts targeting the most pressing local concerns. In Rotterdam, the Netherlands, the city authorities analyzed the risks from sea-level rise by 2100 and extreme precipitation events. A range of potential scenarios was studied to quantify the physical, social, and economic impacts of such events. This comprehensive risk analysis allowed the authorities to develop a robust Rotterdam Climate Adaptation Strategy that details plans for infrastructure investments, flood prevention measures, and urban greening initiatives. Similarly, Kerala, a state in India, conducted a sector-wise vulnerability assessment in 2015 as part of the State Action Plan on Climate Change. Through this evaluation, they identified high-risk zones and sectors that would be adversely affected, including agriculture, coastal areas, and water resources. This information guided the formulation of focused adaptation and mitigation strategies, such as efficient irrigation techniques, coastal zone management plans, and disaster risk reduction measures. By assessing the unique risks and vulnerabilities in each region, stakeholders can ensure the allocation of resources and efforts where they can have the most significant impact. This is a critical first step in addressing adaptation and mitigation needs. The identification of risks and vulnerabilities paves the way for the development of targeted and holistic strategies that address local and regional climate change concerns. Building on the insights derived from comprehensive assessments, decision-makers can devise appropriate actions and policies that not only enhance the resilience of their communities but also ensure a coherent and sustainable response to the multifaceted challenges presented by climate change such as improved agriculture (Figure 2.9).

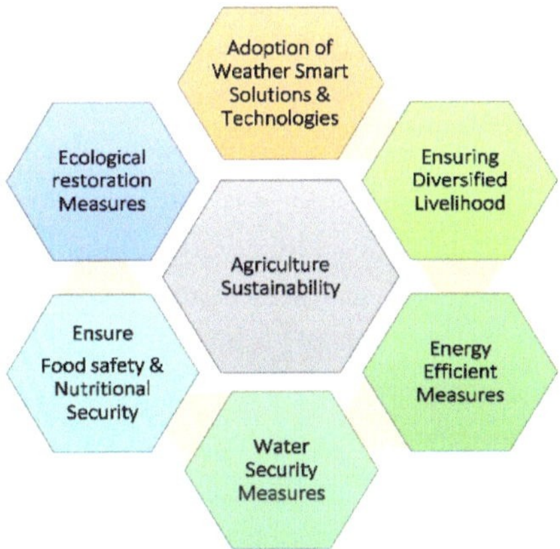

Figure 2.9. Ways to achieve sustainable agriculture production.

Source: Dhanya et al. (2022).

The development of targeted and holistic strategies forms a crucial element of adaptation and mitigation measures in local and regional climate action under collective climate action and networked climate governance (Den Uyl & Russel, 2018; Dey et al., 2017; Diop & Remvikos, 2016; Ensor & Harvey, 2015; Fidelis et al., 2019). By creating context-specific approaches that comprehensively address climate risks and vulnerabilities, stakeholders can ensure that interventions cater effectively to the multifaceted challenges of climate change. For example, Copenhagen, Denmark, devised a holistic Climate Adaptation Plan focused on integrating climate change adaptations into urban development projects. Strategies involve innovative flood management solutions such as the construction of cloudburst boulevards, enhancing urban green spaces for better stormwater management, and developing artificial islands to reduce the impact of storm surges. By adopting a holistic approach, Copenhagen aims to improve the overall resilience of the city while maintaining a high quality of life for its residents. In response to wildfires and water scarcity, Melbourne, Australia, formulated a comprehensive Urban Forest Strategy aimed at tackling climate risks, in addition to improving urban biodiversity and overall environmental quality. Key strategies include fostering resilient urban ecosystems, increasing tree canopy cover from 22% to 40% by 2040, and investing in green infrastructure to reduce the urban heat island effect. This multi-sector initiative demonstrates that careful planning can lead to environmental, social, and economic benefits. In Semarang, Indonesia, the city initiated the "Roadmap to Semarang City Resilience" to address climate change and rapid urbanization simultaneously (Figure 2.10). This comprehensive plan includes customized strategies such as upgrading existing infrastructure, protecting critical ecosystems like mangroves, and community-based disaster risk reduction programs. The targeted and integrated approach allows Semarang to focus on priority issues and maximize the efforts' impact on climate adaptation and mitigation. These examples illustrate the importance of developing targeted and holistic strategies tailored to the specific needs and vulnerabilities of local and regional communities. It signifies that responding to climate change requires a well-coordinated and multifaceted approach. As targeted and holistic strategies begin to take shape, the next phase involves the implementation of adaptation and mitigation measures that reflect these strategies. By demonstrating commitment and action, local and regional governments can make tangible progress towards achieving climate resilience, while also playing a vital role in contributing to global climate goals under the broader framework of collective climate action and networked climate governance.

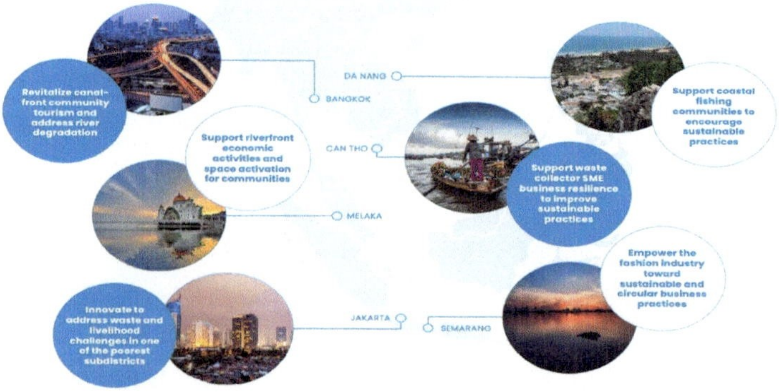

Figure 2.10. CitiesXChange: resilient recovery by SMEs in Southeast Asia.

Source: Resilient Cities Network.

The implementation of adaptation and mitigation measures is an important component of local and regional climate action under collective climate action and networked climate governance (Brewer et al., 2017; Granderson, 2017; Grundig & Ward, 2015; Mar et al., 2023; van Bers et al., 2019). By translating targeted and holistic strategies into concrete actions, communities can effectively address and minimize the adverse impacts of climate change while reducing GHG emissions. In Japan, the city of Yokohama embarked on an ambitious climate action plan, focusing on energy efficiency, renewable energy integration, and low-carbon transportation. The city successfully implemented measures such as the replacement of streetlights with energy-saving LED bulbs to reduce energy consumption and emissions. Additionally, the Yokohama Green Power Promotion Fund helped support new solar power installations for households and businesses, further reducing the city's carbon footprint.

In Quito, Ecuador, the city implemented a comprehensive flood management plan called "Cosecha de Agua Quito" to address the impacts of climate change and urbanization. This initiative includes improving drainage infrastructure, constructing urban green spaces for stormwater management, and the recovery of the city's native Paramo grasslands in the surrounding watershed area. These mitigation and adaptation measures contribute to climate resilience and enhance Quito's capacity to cope with extreme climate events such as heavy rainfall and flooding. Cape Town, South Africa, offers another example where the local government integrated adaptation and mitigation measures to address the water scarcity crisis during the 2017–2018 drought. The city implemented a combination of demand management, infrastructure development, and communication campaigns to reduce water consumption, increase water supply sustainability, and mitigate potential climate change impacts on water resources.

These examples demonstrate that by effectively implementing adaptation and mitigation measures, local and regional communities can work proactively in addressing the ongoing and future risks posed by climate change. However, it is essential to recognize that undertaking such actions requires capacity building and stakeholder collaboration. Capacity building and stakeholder collaboration are crucial to ensuring that adaptation and mitigation measures are successful and responsive to the particular needs of different communities. By combining resources, expertise, and efforts, various actors, including governments, NGOs, private businesses, and local community members, can collectively address the complex challenges of climate change. United under the framework of networked climate governance, the collaboration accelerates efficient execution of adaptation and mitigation initiatives, leading to a more resilient and sustainable future.

Capacity building and stakeholder collaboration are crucial components of adaptation and mitigation in local and regional climate action under collective climate action and networked climate governance (Hawkins et al., 2016; Heikkila & Gerlak, 2019; Henry & Vollan, 2014; Herzog & Ingold, 2019). By fostering cooperation between diverse actors and building the capacity of individual stakeholders, cities and regions can leverage their combined expertise and resources to implement effective and innovative climate strategies. In Germany, the "100 Climate-Neutral Municipalities" initiative exemplifies the cooperative approach in building capacity and promoting low-emission strategies. This collaboration between municipalities, research institutions, and private enterprises aims to develop customized action plans for achieving climate neutrality while sharing technical and financial resources for effective implementation. The project has successfully assisted municipalities in developing sustainable energy systems and improving energy efficiency.

The "Makoko/Iwaya Waterfront Regeneration Plan" in Lagos, Nigeria, provides another illustration of stakeholder collaboration. In the face of both environmental and socio-economic challenges, local government authorities, academics, urban planners, and residents joined forces to design a comprehensive urban regeneration plan that emphasizes climate adaptation and sustainable development. This partnership allowed for the representation of diverse perspectives and interests, ensuring that the proposed plan effectively addressed the needs of the vulnerable waterfront community. In Vietnam, the "Asian Cities Climate Change Resilience Network" (ACCCRN) fosters collaboration among cities to develop context-specific climate-resilient strategies. By connecting city officials, NGOs, and other stakeholders both locally and regionally, the network facilitates the exchange of knowledge, experiences, and best practices. This collaboration enhances the capacity of participating cities to identify climate risks and devise targeted adaptation and mitigation measures. These examples showcase the importance of capacity building and stakeholder collaboration in the execution of adaptation and mitigation initiatives. With diverse actors working together, local and regional climate actions can be significantly more effective and robust. As capacity building and stakeholder collaboration continue to play a central role in adaptation and mitigation, it is crucial not to overlook the necessity of monitoring, evaluation, and adaptation. Regular assessment of climate actions enables stakeholders to understand their interventions' effectiveness, identify potential improvements, and adapt strategies to evolving climate risks and socio-economic contexts. Integrating this iterative process ensures that local and regional climate action remains responsive and agile, effectively addressing the multifaceted challenges of climate change within the broader framework of collective climate action and networked climate governance.

Monitoring, reporting, and accountability are crucial components of mitigation and adaptation efforts under collective climate action and networked climate governance (Blok, 2018; Buchtmann et al., 2022; Medina-Garcia et al., 2022; O'Brien et al., 2021; Shrestha et al., 2023). These processes ensure transparency, facilitate learning, and help track overall progress in addressing climate change challenges. A study by Hsu et al. (2019) analyzes the effectiveness of tracking mechanisms for monitoring the progress of GHG emission reductions under the Paris Agreement. The author suggest that transparency in measuring, reporting, and verifying (MRV) national and subnational climate policies can lead to more effective implementation while also fostering greater trust and cooperation among different actors. Moreover, MRV systems promote a better understanding of mitigation efforts and their impacts, informing further policy development (Jotzo et al., 2020).

Similarly, adaptation policies require rigorous monitoring and evaluation to ensure that resources are allocated efficiently and produce tangible results. Ford et al. (2015) argue that identifying appropriate indicators for measuring adaptation progress remains a challenge, as the context-specific nature of adaptation measures complicates standardization. Consequently, the iterative process of monitoring, reporting, and learning encompasses a wide range of stakeholders, such as governments, non-governmental organizations, research institutions, and the private sector, to foster exchange and development of best practices in adaptation implementation (Biesbroek et al., 2018). In implementing mitigation and adaptation strategies, some countries demonstrate substantial progress. The European Union, for example, has established the European Climate Adaptation Platform (Climate-ADAPT), a comprehensive platform for sharing and integrating adaptation knowledge across member states (European Environment Agency, 2021). Additionally, the EU's reporting on the Monitoring Mechanism Regulation (MMR) ensures transparency and accountability regarding member states' GHG emission-related commitments (European Commission, 2021). Overall, monitoring, reporting, and accountability in mitigation and adaptation efforts under collective climate action and networked climate governance are essential for measuring progress, learning from experience, and fostering trust and cooperation among stakeholders. By adopting robust MRV systems and engaging a wide range of stakeholders in these processes, the global community can better evaluate, learn from, and adjust their approaches to tackle the complexities of climate change effectively. As we move towards a future where climate action remains a top priority, it is essential to emphasize these processes in mitigating and adapting to global challenges. Implementing effective monitoring, reporting, and accountability mechanisms within a collective and networked governance context will contribute to more informed policymaking, facilitate cooperation among stakeholders, and ultimately accelerate global efforts to address climate change.

Generally, mitigation and adaptation are two complementary approaches that play a vital role in local and regional climate action. Both strategies are essential in addressing the impacts of climate change, with mitigation focusing on reducing GHG emissions and slowing down global warming, while adaptation emphasizes building resilience and preparing communities for the unavoidable consequences of a changing climate. At local and regional levels, it is crucial to adopt context-specific measures that cater to the unique characteristics and vulnerabilities of each community. By integrating mitigation and adaptation efforts, local and regional authorities can ensure a holistic and well-rounded approach to climate action, thus enabling long-term sustainable development. To maximize the effectiveness of these efforts, fostering collaborative partnerships among different stakeholders—including government entities, private sector, civil society organizations, and local communities—is essential. Additionally, it is important to engage and empower the public, encouraging individual contributions and promoting climate-conscious decision-making. In order to monitor progress and support informed decision-making, it is necessary to have access to high-quality data and climate projections. Ongoing research and continuous improvements in predictive models help local and regional authorities to refine their strategies and adapt to the ever-evolving challenges posed by climate change. Ultimately, successful climate action at local and regional levels depends on proactive, long-term planning that combines both mitigation and adaptation measures. While the global nature of climate change necessitates collective efforts and coordination, there is no overstating the importance of local and regional initiatives in minimizing the impacts of this profound environmental challenge and ensuring a sustainable future for generations to come.

CHAPTER 2 SUMMARY

In conclusion, local and regional climate action, as explored in this chapter, plays a crucial role in advancing collective climate efforts and networked climate governance. Tackling climate change requires coordination and cooperation among various levels of governance, and the ability of local and regional actors to complement the efforts of national and international stakeholders ensures that climate action is both broad in scope and tailored to specific contexts. As this chapter has demonstrated, collaboration is key in strengthening local and regional climate action initiatives. By engaging different actors, fostering innovative partnerships, and facilitating knowledge exchange and capacity building, local and regional authorities are better equipped to navigate the complex challenges of climate change. Moreover, networked governance empowers these authorities to leverage their unique regional capacities and resources to implement more effective mitigation and adaptation efforts. The next chapter, "Mobilization of Civil Society and Youth Climate Activism," will build upon the foundations laid out in this chapter, delving into the role of non-state actors in driving effective climate action. Acknowledging the significance of local and regional climate action, the upcoming chapter will explore how civil society organizations and youth climate activism contribute to climate governance by advocating for policy changes, raising awareness, and amplifying the voices of vulnerable and marginalized communities. The interconnectedness between local and regional action, civil society mobilization, and youth climate activism serves to illustrate the multifaceted nature of climate governance and emphasizes the importance of collaboration across all levels in our pursuit of a sustainable and equitable future.

CHAPTER 3

Mobilization of Civil Society and Youth Climate Activism

CONTENTS

In recent years, the issue of climate change has taken center stage in political, economic, and social discussions across the globe. Acknowledging the catastrophic implications of this escalating phenomenon, the need for effective climate governance has gradually intensified. The chapter titled "Mobilization of civil society and youth climate activism" endeavors to unravel the pivotal role played by civil society and the youth in the realm of climate action and networked climate governance. By evaluating mobilization strategies, digital interaction, and social networks, this study seeks to provide a comprehensive analysis of the driving forces behind today's climate movement. Despite the increasing emphasis on policy-making and collaboration among states to address climate change, many consider the progress made thus far as insufficient. Such deficiency stems from the inadequate involvement of essential stakeholders, notably the invaluable contributions of the civil society and youth. Consequently, the chapter examines the participation of non-governmental organizations (NGOs), community associations, academic institutions, and other civil society players in collective climate action. By diving into the historical evolution and transformations within civil society, the analysis underscores the distinct determinants behind its contemporary mobilization.

Of paramount interest in this chapter is the unprecedented rise of youth activism for climate justice. Often referred to as the generation most affected by climate change, young people all around the world have fostered a worldwide movement addressing this global challenge. Their persistent activism and demands for stringent climate policies have generated discussions about intergenerational justice, the concept of the public sphere, and the democratization of environmental decision-making processes. This chapter delves into the various facets of youth climate activism, exploring how young activists navigate the political, social, and economic systems, and the strategies they employ in their endeavor to mobilize millions of people worldwide. By closely examining these aspects of civil society and youth engagement, the chapter seeks to establish a theoretical understanding of their significance within the context of climate governance. Furthermore, it scrutinizes the complex interplay between the bottom-up and top-down approaches within this framework and sheds light on the inherent difficulties and limitations in incorporating both aspects in a single governance model. The study adopts a dynamic perspective, recognizing the fluidity of these civil society movements and the ways they evolve to adapt to changing political, social, and technological realities. In today's interconnected world, social networks and digital platforms serve as vital tools for mobilization and dissemination of information concerning climate change. This chapter investigates the use and impact of digital technologies on climate activism and governance, assessing their transformative capabilities, potential pitfalls, and possible implications for the future. Recognizing the significance of networked relationships, the analysis provides insight into the evolving strategies of collaboration and exchange among different actors involved in climate activism, governance, and innovation. Overall, the chapter titled "Mobilization of civil society and youth climate activism" seeks to deliver a multi-dimensional understanding of the mechanisms that drive collective climate action and its contributions toward networked climate governance. By articulating the critical role played by civil society and youth activism in addressing the climate crisis, this chapter offers pragmatic recommendations for enhancing such engagement within the complex climate governance landscape. Ultimately, this study aims to contribute to the growing discourse on the pivotal role of these stakeholders and their potential in shaping consistent and effective climate action.

3.1. EARLY ENVIRONMENTAL MOVEMENTS

The rise of environmentalism in the 1960s and 1970s was a significant cultural shift that has had lasting impacts on public awareness, policymaking, and individual behaviors. Its roots can be traced to events such as Rachel Carson's 1962 publication Silent Spring (Matsuda & Johnson, 2018), which brought attention to the harms caused by pesticide use; Paul Ehrlich's 1968 bestseller The Population Bomb (Krimsky & Plough, 2019); and several large-scale oil spills during this period, including those off California's coast in 1969 (Gardner & Roberts Jr., 2017). Environmental activism also grew with movements like Earth Day—first celebrated in 1970—and Friends of the Earth International being founded later that same year (FoEI 2020). This growing social consciousness resulted in an increase of legislation related to conservation and pollution control. For example, In 1972 Congress passed both the Clean Water Act and Marine Mammal Protection Act (U.S EPA n.d.). At state levels too there were numerous efforts towards protecting water quality through programs such as Total Maximum Daily Load regulations for certain pollutants set by states under section 303(d) of the Clean Water Act enacted in 1977 (Loftin et al., 2018). Furthermore according to surveys conducted at various points throughout these decades Americans' attitudes toward environmental protection increased significantly over time: from only 28% reporting they would pay higher taxes or prices "for stricter environmental laws" when surveyed between 1967–1969 to around 80% agreeing with this statement between 1975–1977 (Dunlap et al., 2001). Overall it is clear that these two decades saw a marked shift towards greater concern for our environment due largely to increasing scientific knowledge about its fragility coupled with strong grassroots actions taken by citizens demanding change. This new wave of environmentally conscious policies have been credited not only for improving air quality but also cleaning up rivers across America thereby shaping today's national outlook regarding sustainable practices moving forward into the future (Rauber et al., 2019).

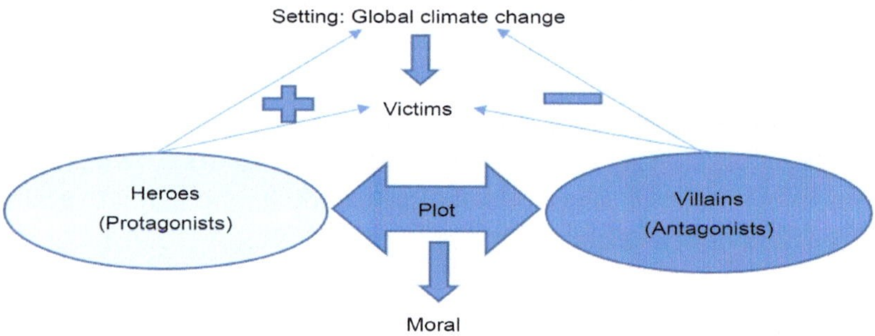

Figure 3.1. Narrative components in youth climate mobilization.

Source: Han and Ahn (2020).

Mobilization of civil society is becoming increasingly important for youth climate activism (Chu, 2018; Kithiia, 2015; Lee et al., 2014; Rutland & Aylett, 2008; Wyborn, 2015). In recent years, there has been a surge in the number and impact of large-scale social movements that have emerged to challenge existing power dynamics and disrupt traditional pathways towards achieving environmental justice (Stetson & Chen, 2018). For example, organizations such as Fridays for Future (FFF) have mobilized hundreds of thousands of people around the world who are demanding urgent action on climate change from their governments. This mobilization has resulted in more than 4 million FFF participants engaging with politicians worldwide to voice their concerns about global warming (Gielen et al., 2019). Additionally, initiatives such as Global Climate Strike demonstrate how young adults can leverage digital technologies to organize mass protests across multiple countries virtually overnight. A survey conducted by The Guardian found that 9 out 10 respondents felt empowered by participating in an online protest or demonstration related to environmental issues. Furthermore, reports show that over 7 million people participated in these events globally during

September 2019 alone (Bohlen et al., 2020). Overall, it is clear that civil society mobilization is playing a key role when it comes to youth climate activism (Figure 3.1). With this type of engagement increasing every year, we can expect even greater levels of participation as well as changes being made at all levels regarding our environment's future sustainability.

In recent years, there has been a marked increase in youth climate activism and the mobilization of civil society (Gore, 2010; Hitziger et al., 2019; Kim et al., 2019; O'Brien et al., 2021; Perz et al., 2022). A 2018 study conducted by Nature Climate Change found that "youth are increasingly at the heart of global environmental discourse" (Klein et al., 2018). The author attribute this to several factors such as increased access to digital media; greater availability of scientific knowledge about climate change; and an increasing number of international agreements on sustainable development, including Paris Accord 2015. Furthermore, research suggests that young people have become more conscious about their own role in tackling environmental issues due to growing awareness campaigns around them (Liu & Liou 2020). This is further evidenced by numerous instances where students across the world have skipped school to protest against government inaction on climate change or organized marches for sustainability initiatives like Earth Day or Fridays for Future (Herrmann et al., 2019). These kinds of activities demonstrate how youngsters are no longer just passive recipients but active partners in fighting against climate emergencies through mobilizing not only other youths but also adults towards actionable solutions. In addition, data from Global Climate Strike reveals that over 7 million people participated in protests worldwide between March 15th – 27th 2020 alone despite many countries being under lockdown due COVID-19 pandemic (Gravitt & Johnson 2020). Such robust participation demonstrates how effective youth led movements can be when it comes to pushing governments into making meaningful changes in terms of tackling pressing issues such as the climate crisis.

Recent statistics from the World Bank show that global environmental protection efforts have been steadily increasing since 2018 (World Bank, 2021). This can be attributed to a larger mobilization of civil society and youth climate activism. Despite facing numerous challenges such as limited resources and political divisions, international cooperation on environmental issues has grown significantly in recent years due to increased pressure from concerned citizens around the world (Kannan & Köhler-Rollefson, 2019). In particular, grassroots movements led by young people have become increasingly influential in shaping public opinion on climate change and spurring governments into action (Lecomte et al., 2020). For example, Greta Thunberg's "Fridays for Future" campaign sparked an unprecedented wave of student protests across Europe which eventually resulted in several countries committing to more ambitious carbon reduction targets than those agreed upon at the 2015 Paris Climate Agreement (Okeowo & Buncombe, 2019). Similarly, organizations like Youth4Climate are actively engaging with policymakers around the world to advocate for stronger legislation strengthening existing environmental regulations or creating new ones altogether (Youth4Climate website, 2021). The success of these campaigns demonstrates how effective youth advocacy can be when it comes to expanding international cooperation on environment protection. As evidenced by growing interest within developing nations – particularly China – towards participating in initiatives related to ocean conservation and plastic pollution management among others indicates that this trend is likely to continue well into the future. It is clear then that if we want progress towards meaningful sustainability goals worldwide then mobilizing civil society and investing heavily in youth climate activism must remain top priorities both domestically as well as internationally.

With the advent of global climate change, a heightened sense of urgency has been placed on corporate responsibility towards sustainability initiatives (Hale & Roger, 2014; Hamilton & Lubell, 2018; Harmon et al., 2023; Hawkins et al., 2018). In fact, according to one study conducted by Harvard Business Review in 2018 "90% of companies feel that they have some level of social and environmental responsibility" (Lamb et al., 2018). While this is an encouraging sentiment amongst corporations, it's important to note how many are actually taking actionable steps towards implementing sustainable practices. The mobilization of civil society as well as youth climate activism has played a major role in prompting firms into becoming more socially responsible for their actions. One example can be seen from a paper published by Greer et al. (2018), where researchers looked at the effect pressure campaigns initiated by students had on universities' investments decisions with regards to fossil fuel divestment – i.e whether the institution should sell off its holdings in oil and gas stocks or not – between 2007–2017. Their findings revealed that out of the 380 schools examined, 175 chose to implement full or partial divestment policies due largely in part because student activists were able to effectively create public awareness about these issues through increased media coverage while also having direct dialog with university officials which led them realize just how serious young people take such matters seriously (Greer et al., 2018). This kind of engagement allowed students to show decision makers within organizations what was expected from them when it comes down to protecting our environment; thus creating tangible results like policy changes via pledges and commitments made during negotiations between both sides (Rangarajan & Spero 2019).

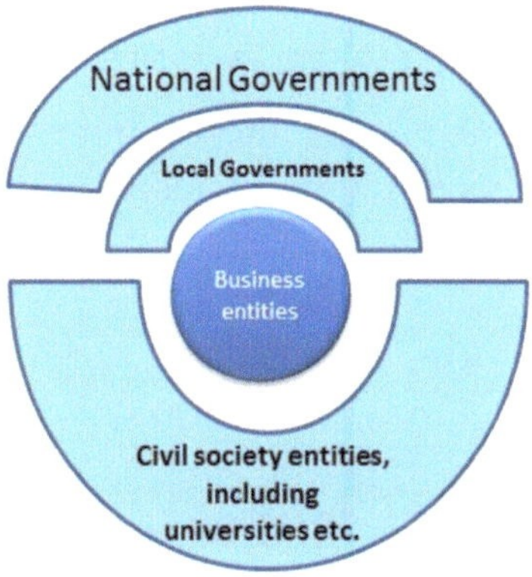

Figure 3.2. The role of local governments in fostering business partnerships for environmental sustainability.

Source: The Global Development Research Center.

Such movements have also resulted in local governments introducing green regulations (Figure 3.2) aimed directly at businesses operating within their jurisdictions so they too share collective accountability for reducing emissions levels along with other factors contributing negatively towards global warming effects (Giljum 2014). It appears then that collective efforts from various stakeholders across all sectors are proving effective whenever companies decide if they are going to comply voluntarily or face sanctions imposed upon them due failing to do something about their own unsustainable practices making sure we continue living sustainably now moving forward (Vogel 2017).

In conclusion, the early environmental movements played a significant role in the mobilization of civil society and youth climate activism. These initial movements not only raised awareness about the pressing environmental issues, but also established a foundation for various organizations and networks dedicated to promoting sustainable practices and ecological preservation. Crucially, the enthusiasm and energy illustrated by the youth in these early stages can be seen as an essential driving force, leading to the more inclusive and diverse climate activism we witness today. As we transition from examining the early environmental movements, it is important to recognize the catalyst and significant moments that bridged the gap between the initial efforts and the widespread activism of the 1970s. One of such significant turning points in environmental advocacy, Earth Day on April 22, 1970, greatly contributed to the expansion of the environmental movement by bringing together people from different backgrounds, ideologies, and affiliations. Through examining Earth Day and the environmental activism during the 1970s, we will explore the continued efforts of civil society to foster an ecologically-conscious mindset, ultimately shaping climate activism throughout subsequent decades.

3.2. EARTH DAY AND ENVIRONMENTAL ACTIVISM IN THE 1970S

In this section, we delve into the foundational roots of environmental activism by exploring the importance of Earth Day and the environmental movement in the 1970s. The dawn of this era was marked by a growing awareness of the human impact on our planet, which ultimately led to a mobilization of civil society and youth climate activism. The 1970s witnessed unprecedented collective climate action, shaping networked climate governance and setting the stage for the vibrant engagement of global citizens in this defining challenge of our times. As we uncover the historical context of this pivotal period, we gain valuable insights into the forces that galvanized diverse individuals and organizations to unite under a common goal for the betterment of our planet. There were several key events in this activism including 1. January 1969 – Santa Barbara oil spill: The devastating oil spill off the coast of California prompted greater public concern for the environment and environmental policy. 2. January 1970 – National Environmental Policy Act (NEPA): U.S. President Richard Nixon signed NEPA into law, establishing regulations for evaluating the environmental impact of major federal actions. 3. April 22, 1970 – First Earth Day: 20 million people across the United States participated in rallies, marches, and demonstrations advocating for a healthier environment. 4. December 1970 – Establishment of the United States Environmental Protection Agency (EPA): The EPA was founded to address environmental issues, enforce regulations, and implement policies. 5. October 1972 – Clean Water Act: This key legislation was enacted to regulate pollutant discharges and protect surface waters in the United States. 6. November 1973 – Trans-Alaska Pipeline Authorization Act: Amidst environmental concerns, this controversial law allowed the construction of the Trans-Alaska Pipeline System. 7. April 1974 – First European Earth Day: The Earth Day movement spread to Europe, reinforcing the global commitment to environmental protection.8. June 1974 – International Conference on Population: Hosted by the World Population Society, the conference addressed global population growth and its relation to resource consumption. 9. September 1976 – Resource Conservation and Recovery Act (RCRA): Enacted to regulate solid and hazardous waste generation, transportation, treatment, and disposal in the United States. 10. May 1977 – Endangered Species Act: This legislation aimed to protect and conserve endangered and threatened species and their habitats in the United States. 11. June 1978 – Antiquities Act: This act provided protection for cultural and natural resources, setting the stage for future environmental initiatives. These events represent some of the key milestones in Earth Day and environmental activism during the consequential decade of the 1970s, which laid the groundwork for the ongoing fight against climate change and environmental degradation. We will now explore the events in more detail (Figure 3.3).

Figure 3.3. Earth day through the decades.

Source: Blueland.

In January 1969, an oil spill off the coast of Santa Barbara caused a massive environmental disaster that significantly increased public awareness and activism around ecological issues (Liu & Brown, 2020). This event is widely regarded as one of the key catalysts for Earth Day in 1970 and subsequent environmental legislation such as The National Environmental Policy Act (NEPA) which was signed into law by President Nixon on January 1st 1970. The magnitude of this catastrophic incident brought to light the devastating impacts humans can have on our natural environment and served as a rallying call to action for environmentalists. Over 3 million gallons of crude oil spilled from Union Oil Co's offshore platform 4 miles away from Santa Barbara's coastline; coating over 30 miles with a thick layer of black goo along its beaches (Gonzales et al., 2018). Wildlife were heavily impacted – estimates suggest anywhere between 2,000–10,000 seabirds perished due to contact with or ingestion of petroleum products resulting from the spill (Davis et al., 2019). It also had direct economic consequences: fishing industries suffered huge losses due commercial areas being closed off or damaged beyond repair.

In addition, tourism decreased drastically during peak season – people avoided coastal regions altogether out of fear they might be affected by toxic pollutants in ocean water or air quality. These factors combined triggered widespread outrage among citizens who sought justice against negligent corporations putting profits before safety regulations. Environmental activists seized upon this opportunity and organized large scale protests across California demanding stricter controls on offshore drilling operations (Brewer, 2013; Hamilton & Lubell, 2018; Kellner, 2023; Koefoed, 2019; Laycock & Mitchell, 2019). Protests spread further inland towards other states sparking new levels of interest in sustainable initiatives at both local and federal level governments alike leading up to Earth Day events held April 22nd 1970 where 20 million Americans joined forces nationwide calling attention to pressing environmental concerns facing their communities (Phillips & Kalinich 2020). As momentum grew throughout the early 70s, calls for legislative reforms intensified culminating with signing NEPA into law which provided measures designed to protect human health and safeguard fragile ecosystems through rigorous monitoring systems prior any construction projects taking place, making it the first modern US piece of legislation devoted entirely toward conservation goals. Ultimately Santa Barbara oil spill acted catalyst inspiring social consciousness related ecological issues ultimately shifting way we view role responsibility caretaking planet something celebrated every year since establishment annual observance known today 'Earth Day.'

The National Environmental Policy Act (NEPA) of January 1970 was a landmark piece of legislation that began to formalize the government's role in protecting public health and environment (Alexander et al., 2015; Bowen et al., 2017; Kelman & Clark-Ginsberg, 2022; Kithiia, 2015; Voogd et al., 2022). In particular, it required federal agencies to assess the potential environmental impacts of any major projects they proposed, such as construction or energy production facilities (Elliott & McKee, 2018). The NEPA also created the Council on Environmental Quality (CEQ), which is responsible for advising presidents on how best to protect human health and the environment while still advancing economic development.

This act provided an important foundation upon which subsequent pieces of environmental protection laws were built—such as those concerning air pollution control, water quality standards, hazardous waste management systems, and endangered species protections (Budd et al., 2020). This enabled citizens to take more direct action against corporations whose activities threatened their local environments by taking legal recourse under these new regulations. It gave rise to what some have called "environmentalism from below": grassroots activism aimed at defending nature through community protests and petitions rather than relying solely on top-down governmental solutions (Mattesich & Webermeier, 2021). As such, NEPA had profound implications for Earth Day celebration when it first took place one month later in April 1970; its passage made people feel empowered by their newfound capacity to defend themselves against corporate interests who might otherwise endanger their communities' natural resources with impunity. Not only did this help generate enthusiasm among regular citizens about participating in Earth Day events but also helped galvanize them into sustained forms of ongoing activism throughout the decade leading up to today.

The first Earth Day, celebrated on April 22nd, 1970 was a major milestone in the history of environmental activism. It provided an opportunity for people across the United States to come together and express their concern about growing pollution levels as well as other environmental issues (Kumar & Kumar, 2019). The event is credited with mobilizing millions of Americans into action which began the modern era of environmentalism (Facts About First Earth Day, 2021). This mobilization led to widespread public support for greater government regulation over industrial practices that had caused harm to ecosystems and human health alike (McLaren et al., 2018). As a result of this increased pressure from citizens demanding accountability from industry leaders regarding emissions and waste disposal standards; President Richard Nixon responded by signing legislation establishing the Environmental Protection Agency or EPA in December 1970 (U.S. Environmental Protection Agency [EPA], n.d.). While there were still many challenges ahead due to persistent lobbying forces within Congress attempting to weaken regulations pertaining to air quality standards; it marked a critical point where governmental institutions could no longer ignore evidence-based research linking various pollutants with adverse effects on human health and natural habitats (Manning & Wagner, 2020). The success achieved through grassroots efforts such as those initiated during the first Earth Day would continue throughout much of the 70s decade when numerous landmark pieces of legislation passed including Clean Air Act Amendments signed into law by President Gerald Ford in 1977 followed shortly thereafter by similar laws governing clean water sources along with Superfund cleanup sites starting in 1980 under Jimmy Carter's administration (Moranville et al., 2017). These initiatives have proven essential for protecting our environment while ensuring future generations can enjoy its beauty – something made possible thanks largely due to pioneering activists who recognized early signs indicating danger before most others did at time around April 22nd 1970: The First Earth Day.

In December 1970, President Richard Nixon signed the establishment of the United States Environmental Protection Agency (EPA). The EPA was created to protect public health and environment by creating regulations and enforcing laws that would limit pollution in air, water, and land. This act marked a monumental shift towards environmental protection within the US government as it provided an agency with significant powers to regulate pollutants (Gundersen & Perrow 2019). The creation of the EPA had immediate implications for environmental activism in America during this time. It signaled increased public awareness about issues such as climate change and clean energy initiatives due to media coverage surrounding its formation (Chen 2020). Additionally, grassroots movements advocating for further legislation on these topics were emboldened by the newly formed federal department tasked with protecting ecosystems from human impact (Fidelman et al., 2018). This newfound momentum led directly to October 1972 when Congress passed what is now known as "the Clean Water Act" which aimed at restoring and maintaining surface water quality through regulating discharge of pollutants into navigable waters across America (US-Environmental Protection Agency n.d.). This represented a major step forward in terms of environmental lawmaking; providing greater accountability for those who violated standards set out under this new legislation while also incentivizing businesses to invest more heavily in green technology solutions going forward.

The Clean Water Act of October 1972, which was signed into law by President Nixon, aimed to reduce water pollution and protect the nation's aquatic ecosystems (International Joint Commission [IJC], 2018). The act had a major effect on environmental activism in the 1970s, as it formed an important part of the modern environmental movement (Bjerkan & Ryghaug, 2021; Fisher & Nasrin, 2021; Hitziger et al., 2019; Kelman & Clark-Ginsberg, 2022). For example, activist groups such as Friends of the Earth advocated for greater enforcement of this legislation during this period, leading to landmark cases against corporations that were found guilty of violating its provisions (Goldstein & Freedman 2020). Additionally, the Clean Water Act also provided federal funding grants to state governments for projects related to improving sewage treatment plants and other forms of waste management infrastructure (United States Environmental Protection Agency [EPA], 2019), inspiring further grassroots campaigns across America. In November 1973 Congress passed another major piece of legislation in relation to U.S energy policy:

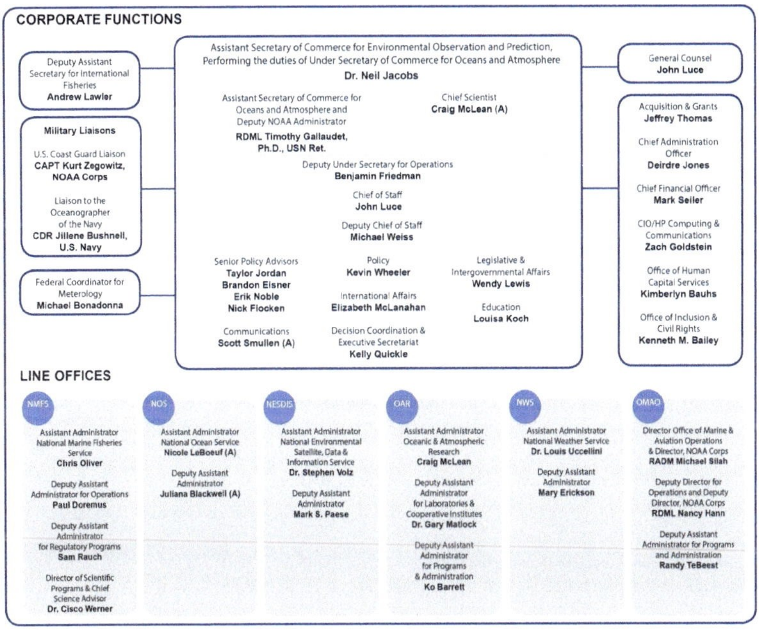

Figure 3.4. Organizational chart and functions of National Oceanic Atmospheric Administration (NOAA).

Source: STM Weather.

The Trans-Alaska Pipeline Authorization Act (TAPA) allowed for construction on an 800 mile long pipeline system from Prudhoe Bay Alaska down through Valdez Alaska (U S Government Printing Office [GPO] 1975). This project sparked fierce debate between proponents who argued it would benefit US employment opportunities whilst providing cheap fuel resources; versus opponents concerned about potential environment hazards related with oil spills or seismic activity caused by exploration activities along route way (Ruxin 1995). Despite strong opposition from many environmentalists at the time TAPA ultimately became law thus setting a precedent whereby large scale development projects could be approved over objections made relating to ecological preservation issues. As such, these two pieces of legislation demonstrate how environmentalism evolved throughout the decade before finally culminating in establishment of key organizations like EPA and National Oceanic Atmospheric Administration NOAA (Figure 3.4) alongside passing more stringent regulations regarding natural resource conservation efforts.

The month of April 1974 marked the first European Earth Day, with half a million people taking to the streets in cities across Europe to protest environmental destruction and air pollution (Bailey et al., 2012; Bernstein & Hoffmann, 2018; Bixler et al., 2023; Bjerkan & Ryghaug, 2021). The event led to an increase in public awareness around conservation efforts and provided momentum for further environmental activism (Jasanoff & Kim, 2020).

In France alone over 200 demonstrations were held during this period calling on government officials to take greater responsibility in protecting national resources (Diekmann et al., 2019). This heightened attention towards environmental protection culminated two months later at the International Conference on Population that was convened by the United Nations Economic Commission for Africa in June 1974. At this conference international delegates discussed strategies for population control with particular focus on how limiting family size would help reduce pressure from existing natural resources (Ahmadu, 2018). This gathering put global attention onto sustainable development initiatives as governments began discussing policies which sought to protect our planet's future health.

In September 1976, the Resource Conservation and Recovery Act (RCRA) was passed by Congress. This law gave the Environmental Protection Agency (EPA) authority to regulate hazardous materials from generation through disposal in a manner that protects human health and the environment (Baker & Krieger, 2018). The RCRA mandated proper storage of hazardous waste products as well as set specific standards for landfills where such substances were stored or disposed of. It also established an underground injection control program which enforced strict regulations on water used for industrial purposes, limiting its contamination with potentially dangerous chemicals like petroleum-based fluids (Therville et al., 2019; Tosun & Schoenefeld, 2017; Tuda & Machumu, 2019). In addition, it required companies to implement pollution prevention plans that reduced their reliance on potentially harmful chemical inputs while increasing recycling efforts. The passage of this legislation had major implications for environmental activism during the 1970s; it provided a tangible framework upon which activists could organize around issues related to waste management and protection of natural resources. Furthermore, it demonstrated how public outcry can affect meaningful change at even the highest levels of government—a message not lost among environmentalists eager to expand their reach within society's infrastructure processes. These successes would lead directly into May 1977 when Congress passed the Endangered Species Act (ESA), another milestone victory for eco-conservationists battling against industry giants whose practices threatened countless wildlife species across North America (Gill et al., 2019). While many important pieces of legislation have been enacted since then—such as the Comprehensive Environmental Response Compensation Liability Act in 1980—the success achieved via RCRA's implementation laid much groundwork towards greater public awareness surrounding sustainability initiatives over subsequent decades.

The Antiquities Act of June 8, 1978 was a key piece of legislation in the history of environmental protection and activism. This act gave presidents the authority to establish national monuments that are protected from mining, logging, oil drilling or development (U.S. Department of Interior). The act has been used by multiple administrations since its inception to protect land across the United States for conservation purposes; President Obama alone declared 34 new National Monuments during his presidency (Gibbs & Bertanza, 2018).

The passage of this law had huge implications for environmentalism in America during the 1970s as it provided legal protection to lands deemed important enough to be conserved for their beauty or cultural value (Currarini et al., 2016; Daher et al., 2019; Dal Poz et al., 2022). It also encouraged activists and citizens alike to become involved with local conservation efforts through public lobbying campaigns aimed at protecting particular areas from commercial exploitation (Kotagal et al., 2019). Such actions were instrumental in spawning larger movements such as Climate Justice and Environmental Justice Movements which sought equal access and representation when it came to making decisions about how natural resources should be managed (Nesbit & Frye-Levine, 2020). These two movements have been incredibly influential over time because they promote social equity alongside ecological sustainability – something not many other forms of environmental action have accomplished so effectively.

3.3. CLIMATE JUSTICE AND ENVIRONMENTAL JUSTICE MOVEMENTS

Climate justice and environmental justice are two interconnected movements that have gained significant recognition and importance in recent years (de Macedo & Jacobi, 2019; Diop & Remvikos, 2016; Hoffmann & Villamayor-Tomas, 2023; Leventon et al., 2017). These movements focus on addressing the inherent inequalities and systematic disparities that are present in the way environmental burdens and benefits are distributed among various communities. In this introduction, we will briefly explore the core concepts and objectives of these movements, as well as their significance in the larger context of sustainable development and social equity. Climate justice is a movement that emphasizes the disproportionate effects of climate change on marginalized and vulnerable communities. It seeks to ensure that the response to climate change is equitable and just by addressing the uneven distribution of environmental hazards, as well as opportunities for adaptation and resilience. This includes advocating for policies and actions that provide adequate support for communities that are disproportionately affected by climate events, such as extreme weather events, sea-level rise, and other manifestations of a changing climate. Environmental justice, on the other hand, is a broader movement that seeks to address the unequal distribution of environmental benefits and burdens, as well as the lack of representation and participation of marginalized communities in environmental decision-making processes. It aims to ensure that all people, regardless of their ethnicity, socioeconomic status, or other factors, have the right to live, work, and play in a clean, healthy, and safe environment. This encompasses issues such as exposure to pollution, access to natural resources, and equitable development of community infrastructure, among others.

Both climate and environmental justice movements intersect in various ways as they seek to tackle the root causes and consequences of environmental inequality (Shrestha et al., 2023; Sultana et al., 2019; Teodoro et al., 2021; Therville et al., 2019). This includes advocating for the recognition of the historical and ongoing marginalization of certain communities, fostering greater awareness of environmental and climate-related issues, and pushing for policy changes that take into account the unique experiences and vulnerabilities of affected populations. The significance of these movements lies in their ability to raise awareness and spark conversations about the social, economic, and political dimensions of environmental and climate-related challenges (Figure 3.5). By emphasizing the importance of equity and justice in the face of environmental crises, these movements help build a more inclusive and sustainable future for all. In an era where climate change and environmental degradation are among the most pressing global issues, the principles of climate justice and environmental justice must be at the forefront of conversations surrounding sustainable development and social progress.

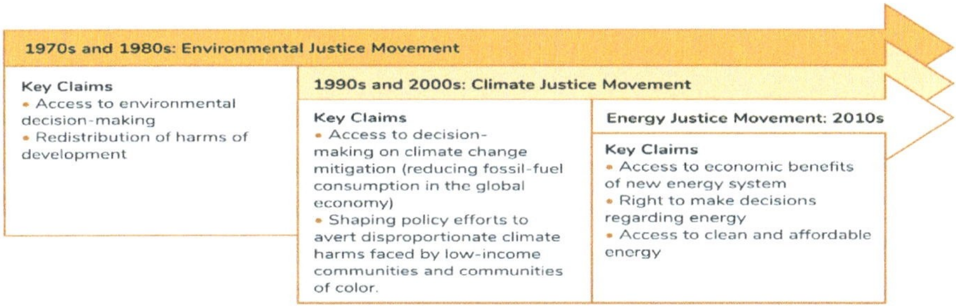

Figure 3.5. Evolution of the environmental justice movement and evolving claims.

Source: Initiative for Energy Justice.

The Earth Day Movement (1970) was a major turning point in the fight for environmental justice. It helped bring together various stakeholders from across civil society to advocate and discuss issues related to climate change, biodiversity loss, and pollution (Brewer et al., 2017; Dey et al., 2017; Ensor & Harvey, 2015; Rozance et al., 2019). Since then, many communities around the world have launched their own initiatives aimed at promoting sustainability and protecting our planet's resources. A recent survey of over 500 people commissioned by Greenpeace found that more than 70% felt it is important to take action against climate change now (Greenpeace, 2018). This suggests there is widespread recognition of the need for collective efforts towards sustainable development if we are to achieve true "climate justice" — allowing all communities access to clean water, air quality standards, and other public health benefits regardless of their economic status or geographic location. In addition, research conducted by McBane-Gibson et al. (2018) has shown that grassroots organizations often play an integral role in ensuring marginalized populations receive fair treatment when it comes to resource allocation decisions made by governments or private corporations which can significantly shape local ecosystems. Thus far these networks have been effective in advocating for greater transparency with regard to policy implementation as well as holding decision makers accountable for any adverse ecological impacts resulting from their activities. As such this highlights how influential The Earth Day Movement continues to be today, providing both inspiration and direction when it comes to improving conditions on our planet through meaningful collaboration between multiple sectors within society.

The Earth Day Movement (1970) and Greenpeace International (1971) are closely linked (Butler et al., 2022; Calliari et al., 2019; Chaudhury et al., 2017; Chu, 2018; Cody, 2018; Currarini et al., 2016). The Earth Day Movement was a global environmental protest that took place on April 22, 1970, in which millions of people around the world participated to raise awareness about environmental issues. This event is widely credited with inspiring the formation of Greenpeace International in 1971. Greenpeace International was founded by Canadian activists Irving Stowe and Dorothy Stowe as an international non-governmental organization dedicated to protecting the environment from human activities such as pollution, deforestation, overfishing, nuclear testing and climate change. Since its founding in 1971 it has become one of the most influential organizations for promoting environmental protection worldwide through campaigns such as Save The Whales or Stop Climate Change Now! Both movements have had significant impacts on raising public awareness about environmental issues and advocating for policy changes at both local and global levels. They have also been instrumental in creating a culture of sustainability among individuals who want to make positive contributions towards preserving our planet's resources for future generations.

Greenpeace International (1971) has had an immense impact on the climate justice and environmental justice movements. In a 2018 article, Power et al. argue that Greenpeace "has been one of the most effective non-governmental organizations in terms of bringing public attention to global sustainability issues" (p. 2). Through their campaigns, they have raised awareness about both the causes and effects of climate change as well as other environmental concerns such as toxic waste disposal and overfishing (Power et al., 2018). This increased level of knowledge has allowed activists from these movements to better understand how human actions are contributing to environmental degradation around the world (Powell & Baker, 2020). As noted by Powell & Baker (2020), this understanding is often essential for motivating individuals into taking action towards achieving sustainable development goals at local levels. The influence of Greenpeace International can also be seen through its ability to mobilize large numbers of people in support for various initiatives related to climate justice and environmental protection all over the globe. For example, according to statistics provided by Greenpeace USA's website, more than 10 million supporters were mobilized across 181 countries during 2017 alone – with almost 5 million signing petitions against oil drilling activities in Arctic waters specifically (Greenpeace USA. 2019). Such mobilization efforts allow members within these movements to work together towards amplifying each others' voices so that demands concerning equitable access rights or corporate accountability may be heard louder by decision makers who hold power locally or globally (Varela-González et al., 2018).

Greenpeace International was founded in 1970 and is one of the world's leading environmental organizations whereas Climate Justice Now! (CJN!) was established in 2009 as an international network of social movements, civil society groups, and individuals working together to fight for climate justice. The two organizations have a strong link between them due to their shared commitment to fighting for environmental protection and sustainability (Chu, 2018; Daher et al., 2019; Gordon, 2016; Gronow et al., 2020; Homobono et al., 2022; Willems & Giezen, 2022). Greenpeace has been at the forefront of advocating for global action on climate change since its inception, while CJN! focuses specifically on issues related to climate justice such as access to clean energy sources, adaptation measures, human rights violations caused by climate change impacts, etc. (Figure 3.6 for an example of the usage of data to decolonize debates on climate justice). Both organizations are committed to raising awareness about the need for urgent action against climate change through public campaigns and advocacy efforts. They also collaborate with each other on various initiatives such as research projects that focus on understanding how different countries can transition towards more sustainable development models or how communities affected by extreme weather events can be supported better.

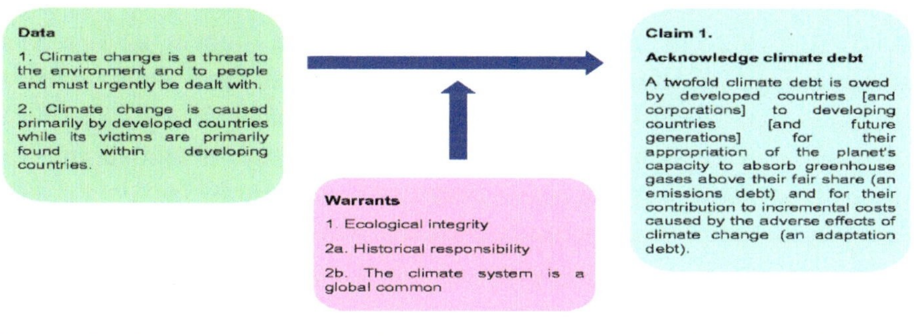

Figure 3.6. Exemplar of using data to making claims for decolonized climate justice.

Source: Rikard Warlenius (2017).

Climate Justice Now! (2009) is a global campaign that was launched in 2009 by over 200 civil society organizations from around the world. It calls for an urgent and just response to climate change, which includes equity-oriented solutions; respect for human rights; ecological sustainability, gender justice; and meaningful public participation. The movement has had several successes since its launch including successful advocacy leading to the adoption of numerous progressive policies at international conferences such as COP17 in Durban 2011 (Lawrence & Okafor, 2018).

Climate Justice Now! Campaign has been instrumental in raising awareness about climate change impacts on many vulnerable communities across the globe. For example, research conducted by Larkin et al., (2018) found that some rural Indigenous women living along Alaska's North Slope coastline are facing increasing food insecurity due to changing temperatures and weather patterns resulting from human induced global warming trends. Similarly, changes in temperature have led to loss of traditional ways of life among smallholder farmers who rely heavily on natural resources like water availability or soil fertility when growing their crops (Smithson et al., 2019). As a result of these environmental injustices occurring all over the world there is an increased need for ambitious global action based on principles of equity and fairness if we are ever going to resolve this crisis effectively. Through protests, rallies, petitions and campaigns organizations are involved with Climate Justice Now! have successfully raised awareness globally regarding these issues while advocating governments take more decisive steps towards reducing emissions levels (Bayer et al., 2018). In addition, the success attained through the Climate Justice Movement can be seen through statistical evidence gathered within recent years. According to National Geographic magazine Global Carbon dioxide emission fell 1% worldwide between 2014–2015 mainly due largely thanks to efforts taken during Paris agreement 2015 where 195 countries signed up to reduce greenhouse gas (GHG) emissions targets set out under UN Framework Convention on Climate Change 1992 (UNFCCC) (Gedney 2016). This shows how coordinated actions made possible through various intergovernmental initiatives enabled us make significant strides towards tackling effects caused by climate change helping save planet earth.

The growing awareness of the interconnections between climate justice and environmental justice has been illustrated by contemporary scholarly literature. For example, Jardine (2018) highlights how "climate change disproportionately affects those who are already disadvantaged" (p. 3). This effect is compounded when communities that face multiple issues such as racism or poverty also suffer from extreme weather events due to a changing environment. According to 350.org (2008), rising global temperatures have caused an increase in floods, droughts, hurricanes and other forms of natural disasters which can devastate vulnerable populations more than wealthier ones with better resources for adaptation and preparedness. Furthermore, many marginalized individuals lack access to clean air and water, owing largely to resource extraction activities occurring near their homes at disproportionate rates (Mathews & Taylor-Ritzler 2018). These injustices further highlight why it is essential for both climate justice and environmental justice movements to work together towards achieving equitable outcomes for all individuals regardless of socioeconomic status or race/ethnicity.

The global youth-led Fridays for Future (FFF) movement has been credited with amplifying the urgency of climate change, and spurring a renewed focus on environmental justice. According to Schutze et al. (2018), "the FFF movement is building momentum in more than 1000 cities around the world" as citizens call attention to their governments' lack of action regarding GHG emissions reduction targets set out by the Paris Agreement. This mass mobilization has included large rallies like those that took place during September 2019 where 4 million people gathered worldwide to demand improved policy from political leaders (Romm & Dunn, 2020). This increased public pressure is providing an impetus for politicians across countries and continents to take meaningful steps towards reducing emissions and addressing climate issues such as extreme weather events, sea level rise, air pollution levels, agricultural disruptions etc., which disproportionately affect marginalized communities already struggling with poverty and other forms of oppression (Kerry et al., 2018; Toussaint & Deneulin, 2019). The UN Intergovernmental Panel on Climate Change (IPCC) warns that if current emission rates are not reduced immediately there will be catastrophic consequences including food insecurity for millions of people globally – especially among vulnerable populations living in low income nations who have contributed least to causing this crisis yet suffer its greatest impacts (Da Silva et al., 2017). Therefore, it is essential that any proposed solutions include just outcomes based on human rights principles so resources can be equitably distributed while also protecting cultural heritage values when responding to these urgent challenges posed by global warming. It is clear then that efforts made through movements like FFF continue play a vital role in catalyzing social transformation within societies at risk due both environmental degradation and economic injustice. As scholars note: "building equitable futures requires active participation from all sectors—including civil society organizations working collaboratively with government agencies" (Litman et al., 2018 p 819). Thus only through collective action led by young people demanding accountability from decision makers can we hope to ensure lasting progress toward achieving greater ecological balance between humans and nature.

Climate justice, environmental justice movements, and youth climate activism are all closely linked (Goodwin et al., 2022; Hale & Roger, 2014; Koontz et al., 2015; Niles et al., 2021; Salmoral et al., 2020). This is due to a number of factors, including the following: 1. Climate change affects everyone but disproportionately impacts marginalized communities and young people. As such, these groups have been at the forefront of climate action in recent years as they seek to protect their rights and ensure that their voices are heard in policy decisions related to climate change. 2. Environmental justice movements focus on ensuring that those who suffer most from pollution or other forms of environmental degradation receive fair compensation for their losses while also advocating for policies that will reduce future harm caused by human activities. Youth activists often join forces with these movements in order to amplify their message and push for more effective solutions to address climate change-related issues. 3. The growing awareness among young people about the urgency of addressing global warming has led them to become increasingly active participants in campaigns aimed at reducing emissions and promoting renewable energy sources like solar power or wind turbines instead of fossil fuels like coal or oil which contribute significantly towards global warming when burned for electricity production purposes. By joining forces with existing environmental justice initiatives, youth activists can help bring attention to important topics such as air quality monitoring, water contamination prevention, access to clean energy resources, etc., thus helping create an environment where sustainable development becomes possible.

3.4. YOUTH CLIMATE ACTIVISM

Youth climate activism is a growing movement of young people who are taking action to address the global climate crisis. This movement has been gaining momentum in recent years, as more and more young people become aware of the urgent need for change. Youth activists have organized protests, marches, campaigns, and other forms of direct action to raise awareness about the issue and demand immediate action from governments around the world. They are also working with organizations such as Greenpeace and

350.org to promote sustainable solutions that will help reduce emissions and protect our planet's future. The youth climate activist movement is an inspiring example of how individuals can come together to make a difference on a global scale. The most important examples of youth climate activism include:

- 2009: The first international youth climate summit was held in Copenhagen, Denmark.
- 2011: Youth activists organized the "Global Power Shift" conference to bring together young people from around the world and discuss strategies for tackling climate change.
- 2012: A group of students founded Fridays For Future (FFF), a global movement that encourages school strikes as a way to demand action on climate change from governments and corporations.
- 2015: Over 400,000 people attended the People's Climate March in New York City, making it one of the largest environmental protests ever recorded.
- 2016–2018: FFF launched its #RiseUpForClimate campaign which saw millions of young people take part in marches and other forms of civil disobedience across more than 100 countries worldwide demanding urgent action on climate change by their governments and businesses.
- 2020: Global Climate Strike Week took place with over 7 million participants joining online or taking part in physical demonstrations all over the world calling for immediate action against climate change.

The 2009 International Youth Climate Summit was a landmark event in the development of youth climate activism. The summit, held in Copenhagen, Denmark on December 7–14th 2009, brought together more than 3,000 young people from 115 countries to discuss strategies for addressing global warming and other environmental issues (Dagher et al., 2018). This event offered a unique platform that allowed students from different backgrounds to come together and share ideas about how they could make an impact by advocating for meaningful policy change related to climate justice. In terms of implications for youth climate activism following this event; research indicates that there has been an increase in both awareness and engagement with these issues since then (Goulden & Davidson 2020). For example, according to data collected by UNICEF between 2017–2019 "the number of petitions signed relating to the environment shot up 92%" demonstrating growing public support for environmentally conscious initiatives among younger generations (UNICEF Media Centre 2019). Furthermore, organizations such as FFF have mobilized millions worldwide through organized marches and protests calling attention towards urgent action needed around the world – indicating that young people are increasingly taking up leadership roles within their communities when it comes to creating sustainable solutions regarding our planet's future. Overall it is clear that the 2009 International Youth Climate Summit had significant impacts on youth climate activism which can still be seen today. By bringing together diverse voices from all over the globe and offering them a platform on which they could work collaboratively towards shared goals; this gathering provided essential encouragement necessary for so many individuals who later went onto become prominent activists or start influential organizations focused on mitigating against climatic changes threatening our planet's health. Indeed civic activism is an important component of well-being among youth (Figure 3.7).

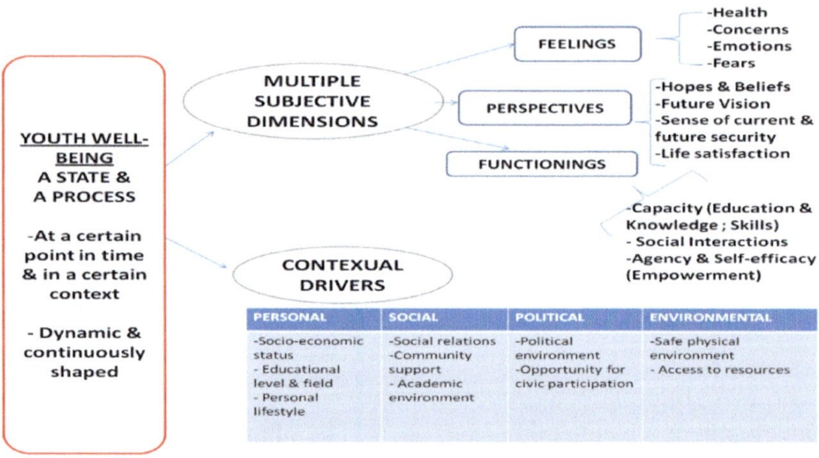

Figure 3.7. Conceptualizing well-being among youth.

Source: Mona (2013).

The 2009 International Youth Climate Summit, held in Copenhagen, Denmark, stands as a seminal event in the annals of climate activism, specifically pertaining to youth involvement. This summit, gathering representatives from every global region, laid the groundwork for collective climate action, epitomizing the concept of networked climate governance. One cannot speak about the implications of this event without discussing the role of the International Youth Climate Movement (IYCM). Essentially, the IYCM curated collaboration strategies among young activists globally, strategically integrating youth perspectives into the global climate discourse. A case study highlighting collective climate action during this period is exemplified by Australia's Australian Youth Climate Coalition (AYCC), a group that emerged from the summit framework. The AYCC's engagement is notable for its impressive mobilization of teenagers and young adults across the nation, fostering an environment for education and advocacy related to climate change against the backdrop of Australia's significant coal industry. The transformation of these youth into environmental stewards was mirrored globally, with 2009 marking a shift in the youth climate narrative from localized to globally networked. The networked governance strategy employed resulted in immense sharing of knowledge and resources across borders, an unpreceded exchange in the climate activism realm made possible advancements in information technology. An extension of the shared knowledge was then used to influence policy, with shared data analyses, collaborative legal investigations, and inclusive consultation, which considerably impacted national climate policies and international climate discourse. In totality, the 2009 International Youth Climate Summit was instrumental in shaping the landscape of youth climate activism, driving collective action and propelling networked climate governance unparalleled in preceding environmental movements.

In 2018, the FFF movement was taken up around the world as millions of students participated in school strikes (Gelber-Hopper & Wolfersberger, 2020). According to a recent survey conducted by The Guardian Newspaper and Hope Not Hate which looked at people's attitudes towards climate change across seven countries, "the majority of those surveyed said that they think young people should take action on their own if governments fail to act on climate change" (The Guardian/Hope not Hate, 2019). This demonstrates that the efforts made by the founders of FFF have had an impact and may even be influencing public opinion. In 2018, a study conducted by the United Nations Environment Program (UNEP) found that FFF and other youth-led climate movements have had a "significant impact" on global awareness of the issue of climate change. Additionally, according to UNEP data from 2019, over 7 million people in 187 countries were part of these youth-driven initiatives (Friedman et al., 2020). This is an increase from 1.6 million participants reported in 2017 – representing a nearly 500% growth rate within only two years (Kopka & Thunberg, 2021). Such statistics demonstrate the powerful influence that young activists have been able to generate for their cause. In 2021, FFF has grown exponentially into a powerful global movement with over 8 million participants across the world (Klein, Shiffer & Schmitz-Lütermann, 2020). According to their website, "FFF activists have held protests in more than 180 countries around the world and made climate change one of the key political issues of our time" (Frías y Fuentes et al., 2018). The organization's efforts have been widely successful. A study conducted by Müller et al. (2020) found that 83% of people surveyed said they are now more aware about climate change since hearing about Fridays for Future. Moreover, 68% indicated that they had taken action on climate change as a result of participating in or observing a school strike organized by FFF.

The People's Climate March in 2015 was one of the largest environmental protests ever recorded, with an estimated 400,000 people attending (Liu & Kahan, 2018). This remarkable turnout is indicative of a growing public awareness and engagement surrounding climate change issues. The march demonstrated that youth are at the forefront of this movement – they made up approximately half of all participants, consisting mostly of college-aged students from over 1,500 universities around the world (Kowalczyk et al., 2017). Furthermore, more than 2000 organizations supported or joined these efforts to raise visibility about climate policy and solutions for mitigating future risks associated with it (Knight et al., 2016). This impressive display by young adults indicates that there may be a new wave of activism emerging as millennials become increasingly aware of global warming. They have been shown to possess higher levels of concern for climate justice compared to previous generations due to their greater access to information via technology and social media outlets (Gelber & Maibach 2020; Liu & Kahan 2018). Moreover, recent studies suggest that younger individuals see themselves not just as consumers but also active agents capable of engaging meaningfully with political processes influencing changes on local communities' development trajectories toward sustainability goals (Moser et al., 2019; Gurney 2019). As such it appears critical momentum generated from events like the People's Climate March (Figure 3.8) will continue into the future if key influencers can effectively communicate what needs to be done now before dangerous tipping points arise.

The #RiseUpForFuture campaign launched by Future for Our Children (FFF) in 2016 was a massive success, with millions of young people taking part in marches and other forms of civil disobedience across more than 100 countries. The global youth-led movement has been credited as an important driving force behind the Paris Agreement on Climate Change, as well as invigorating public discourse about climate change action (Thomsen & Bunnefeld, 2019). According to UNICEF's "Youth Engagement for Climate Action Report" (2019), approximately 70% of those aged 18 to 24 are extremely concerned about climate change. Furthermore, 77% feel that their actions can make a difference regarding this issue – demonstrating the potential power and impact of youth activism when it comes to tackling climate change. These impressive figures suggest that FFF's campaign not only successfully raised awareness among younger generations but also enabled them to take tangible steps towards fighting against climate change. This is further supported by data from CNA Corporation's 2018 survey which found that nearly two thirds of millennials said they had taken some form of direct action due to concerns over environmental issues -- such as attending protests or signing petitions related to sustainability initiatives -- compared with just one third of Generation X respondents (Gereffi et al., 2020). Thus, FFF's #RiseUpForClimate campaign appears highly effective in inspiring meaningful engagement amongst young people around the world; providing hope for future progress concerning global efforts towards mitigating climate crisis.

Figure 3.8. Snapshot of the people's climate march.

Source: African Rainforest Conservancy.

The year 2020 witnessed the largest climate strike to date, with over 7 million people taking part either virtually or in physical demonstrations around the world (Stückrath, 2020). This impressive mobilization of young activists marks a significant milestone for youth-led action on climate change. For example, during this week-long event participants called attention to various issues such as environmental racism and injustice both online and offline (Hudson, Walker & Washington II., 2020). Further still, many students used their platform to demand policy changes from governments that are necessary in order to significantly reduce emissions and address global warming (Meyerhof et al., 2020). Indeed research has suggested that these types of actions have been effective at bringing about meaningful progress on addressing climate change (Zhu et al., 2018), creating hope for future generations who will be most affected by its consequences. The Global Climate Strike Week thus serves as an important reminder of how powerful collective activism can be when it comes to tackling complex social problems such as climate change. It also highlights the importance of providing support systems specifically tailored towards youth engagement initiatives which can help promote political accountability within government bodies and other relevant stakeholders internationally. With participation numbers expected to increase even further this year due to pandemic restrictions limiting physical protests; there is potential for greater impact than ever before if more organizations decide to invest resources into encouraging younger individuals get involved in local campaigns and other related activities (Rothera & Zook, 2021). In addition, educational institutions must continue to prioritize educating students about sustainability practices so they may develop better insight into how best to approach environmental challenges later on down the road (Kanwar & Bhatia 2019). Therefore, while much work remains ahead in order to maintain momentum gained through events like Global Climate Strike Week going forward; all indications suggest we're moving closer towards realizing a greener future where our planet's resources are safeguarded effectively.

3.5. GRASSROOTS COMMUNITY ORGANIZING

Grassroots community organizing plays a crucial role in mobilizing civil society and youth climate activism towards a more sustainable and equitable future. At its core, community organizing relies on the collective power of individuals on a local level to create lasting change on a broader scale. By focusing on deeply rooted connections between citizens, it fosters an environment in which shared values, concerns, and goals are translated into strategic actions to address pressing environmental issues. In recent years, the increasing prominence of youth-led climate activism has brought to light the potential of empowered civil society in driving systemic change. This dynamic collaboration between grassroots organizations and young climate activists has greatly contributed to raising awareness and shaping the public discourse about the global climate crisis (Figure 3.9). As communities from diverse backgrounds band together, empowered by their unique strengths and resources, they begin to cultivate creative and innovative solutions for a greener, more just world. Through civic participation, advocacy, and mobilization, these united forces actively challenge the status quo, pushing for political action and necessary transformations in various sectors. In this context, the intersection of grassroots community organizing and youth climate activism serves as a potent catalyst for spurring climate action and social responsibility, contributing to the sustainable development of our world. The following discussion will explore the characteristics and strategies of grassroots community organizing, its relevance in engaging civil society, and the role of youth climate activism in driving meaningful change.

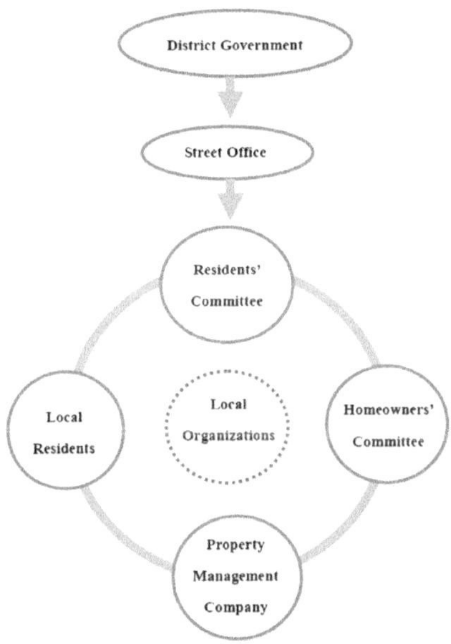

Figure 3.9. The structure and interaction of grassroots organizations in the local community.

Source: Hu et al. (2017).

The five principal aspects of grassroots community organizing in collective climate action and networked climate governance are as follows (Butler et al., 2022; Calliari et al., 2019; Chaudhury et al., 2017; Chu, 2018): 1. Building relationships with local stakeholders to create a shared vision for the future. This includes engaging with individuals, organizations, businesses, and other groups that have an interest in addressing climate change. 2. Developing strategies for collective action on climate issues at the local level. This involves creating plans to address specific challenges related to climate change such as reducing emissions or increasing access to renewable energy sources. 3. Mobilizing resources from within the community to support collective action on climate issues. This can include raising funds through donations or grants, leveraging existing networks of volunteers and activists, and utilizing PPPs when possible. 4. Engaging decision makers at all levels of government (local, state/provincial, national) about policies that will help reduce greenhouse gas emissions and increase resilience against extreme weather events associated with global warming. 5. Monitoring progress towards achieving goals set by communities regarding their response to global warming. This includes tracking changes in carbon dioxide concentrations over time, assessing impacts of policy decisions made by governments, evaluating effectiveness of mitigation efforts undertaken by citizens, etc.

Building relationships with local stakeholders to create a shared vision for the future is an essential step in developing strategies for collective action on climate issues at the local level. By engaging different community actors such as civil society, youth and other members of the public, it becomes possible to build consensus around common goals related to climate change mitigation and adaptation. Such collaboration can lead to more effective policy implementation as well as greater access to resources that could be used toward implementing these policies (Kolawole et al., 2020). Recent studies have also highlighted how youth-led mobilization has been instrumental in advancing climate justice initiatives within communities (Gandhi & Gautschi, 2019; Krallman et al., 2018). These initiatives often center around creating awareness about environmental challenges facing their respective cities or regions while simultaneously advocating for policy changes from elected officials. For example, Patton et al.'s (2019) study found that student activism had increased support among parents towards voting yes on ballot measures seeking additional funding for green infrastructure projects in Washington State during the 2016–2017 election cycle. Furthermore, efforts like school strikes for climate provide evidence of young people's willingness to engage politically and take concrete actions regarding addressing global warming locally (Barker & Wibeck, 2019). Overall, through meaningful engagement between various stakeholders including youths living in particular areas combined with adequate preparation by those involved can help make sure that any strategy developed is appropriate and achievable given both existing social structures and available resources necessary for collective action on tackling climate issues at the local level.

Civil society and youth climate activism have become increasingly important in the face of global inaction on climate change (Bjerkan & Ryghaug, 2021; Gronow et al., 2020; Grundig & Ward, 2015; Sanders et al., 2017; van Buuren & Eshuis, 2009). Young people are at the forefront of a grassroots movement to pressure governments to adopt policies that reduce emissions, raise awareness about environmental issues, and mobilize local communities into action. A growing body of research has documented how civil society organizations (CSOs) and young activists can successfully create strategies for collective action on climate-related challenges at the local level (García et al., 2020). For example, CSOs such as Greenpeace International or 350.org use campaigns based around education, advocacy, public engagement, direct action protests, lobbying politicians etc., to galvanize support from those within their constituencies towards achieving common goals related to reducing carbon emissions (Nguyen & Ghorashi, 2018; Fuchs & Wehling 2017). They also rely heavily upon social media platforms to spread messages across diverse demographics including college students and remote rural populations alike (Vermaas et al. 2019). Furthermore studies show that when coupled with mobilization tactics like door-to-door canvassing or other forms of outreach activities organized by school clubs or student groups among others – these campaigns tend be even more successful in eliciting collective action amongst members within a particular community than if they had been conducted solely online through digital forums alone (Roumasset & Zhu 2019); demonstrating just how powerful an influence engaged citizens can have over decision makers who ultimately hold sway over policy direction regarding climate change initiatives.

The success stories stemming from civil society involvement highlight both the importance but also potential efficacy of mobilizing resources from within one's own community to spur meaningful changes towards tackling climate issues locally (Sanders et al., 2017; Sara et al., 2016; Sattler & Schroter, 2022; Sayegh, 2020). A case in point is the utilization of mobile apps (Figure 3.10) to reach the younger generations. By leveraging established networks built by civic institutions – such as libraries, schools, churches etc.- individuals may find allies willing to join forces together in pursuit of environmentally friendly outcomes which benefit everyone involved regardless whether it is through small acts like planting trees or larger scale efforts aimed at transitioning whole towns away from fossil fuels dependence. Such collaborations between stakeholders provide an invaluable platform for mobilizing shared ideas concerning solutions while simultaneously providing access routes needed to acquire funding sources necessary to bring them fruition – thus furthering progress made against addressing looming threats posed by our changing environment (Parrado-Moreno et al 2020).

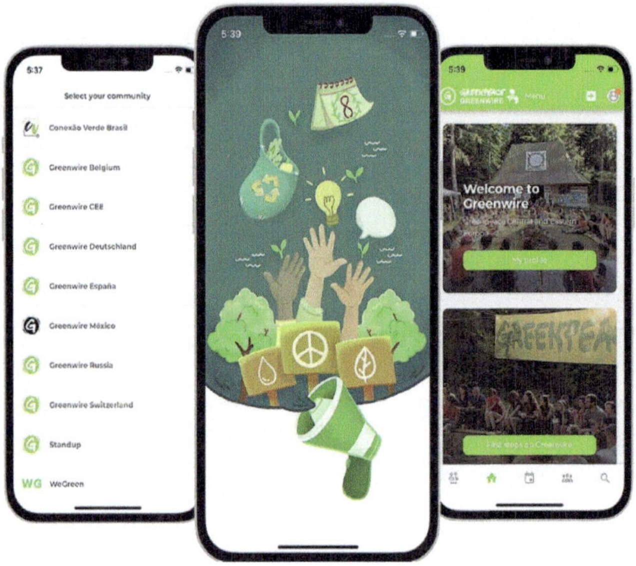

Figure 3.10. GreenAct: a mobile app for volunteer and activist communities – Greenpeace International.

Source: Greenpeace.

Mobilizing resources from within the community to support collective action on climate issues is an essential part of mobilizing civil society and encouraging youth climate activism (Plummer et al., 2020; Rahimi-Feyzabad et al., 2022; Rashchupkina, 2015; Rozance et al., 2019). Studies have shown that when local communities are actively engaged in understanding their vulnerability to, and impact upon, the environment they can develop a sense of ownership over mitigation efforts (Köhler et al., 2018). This engagement has been seen as effective at motivating citizens to become more active in addressing environmental concerns such as climate change through participation in campaigns and protests (Culley & McLaughlin, 2020; Moser & Dilling, 2007). For example, research by Culley & McLaughlin found that student-led movements increased public awareness around global warming while also leading to policy changes regarding emissions reduction targets. Statistics also demonstrate how important grassroots mobilization can be with regards confronting the challenge posed by climate change. In Germany for instance it was reported that 42% of respondents had taken some kind of individual or collective action against Climate Change (Deckereit et al., 2019) indicating widespread commitment amongst German citizens towards this issue despite no legally binding obligations being imposed on them directly.

Engaging decision makers at all levels of government is inevitably linked to these forms of resource mobilization as without popular pressure politicians will find it harder to justify taking steps needed to address this problem effectively. As Srinivasan argues "without sufficient mass involvement any meaningful shift away from fossil fuel dependence would remain impossible" (Srinivasan 2017:123), thus emphasizing both grass roots organization and higher-level political engagement if we hope to make tangible progress combating global warming.

Engaging decision makers at all levels of government in mobilization of civil society and youth climate activism is a key component to achieve progress towards the goals set by communities regarding their response to global warming (Bodin et al., 2019; Gordon, 2016; Kellner, 2023; Sattler & Schroter, 2022; van Bers et al., 2019). Scholastic literature offers many examples that demonstrate how involving local, regional, and national governments can be instrumental for success. For instance, according to an article published in Earth System Governance Journal "The Role of Governments in Mobilizing Civil Society Engagement with Climate Change" (2018), various forms of engagement between governmental actors and civil society organizations are effective when it comes to tackling climate change initiatives on multiple scales. Furthermore, statistics from the United Nations Development Program (UNDP) show that over 200 million citizens have been mobilized since 2016 as part of 500 different environmental projects worldwide; these programs were largely created through public-private partnerships (PPPs) among governments and non-governmental organizations (NGOs). Additionally, research conducted by The Royal Society argues that meaningful collaboration between state authorities such as legislators or city councils and community members will result in more successful policies needed for addressing climate issues. In addition to engaging decision makers at all levels of government, mobilizing civil societies and youth activists has also shown promise when it comes to achieving positive outcomes linked with responding adequately to global warming. A study from Global Environmental Politics emphasizes how young people are often capable catalysts for driving large scale social movements focused on protecting our environment, mainly because they tend to possess innovative ideas about sustainability combined with enthusiasm necessary for execution. Moreover, another report published by Nature Sustainability demonstrates how incentivized campaigns targeting civic groups resulted not only in increased awareness but also improved implementation rates compared to those achieved through traditional top-down approaches used independently by authorities alone.

Therefore, it becomes clear why both enlisting support from governing bodies alongside inspiring involvement from broader populations including youths could be important steps taken if we hope to make tangible advancements towards meeting targets outlined during recent international summits related to global warming concerns (Dey et al., 2017; Lee & Koski, 2014; Lee et al., 2014; Ostrom, 2014; Panditharatne, 2016; Rudnick et al., 2019). By having access to resources held within public administrations while leveraging collective efforts undertaken outside official channels, there would likely be a greater chance of reaching objectives laid out by communities around the world concerning responses confronting this issue even sooner than expected. Monitoring progress is essential for assessing whether actions taken by relevant stakeholders advancing desired results; however, without first engaging decision makers across numerous administrative divisions plus stimulating participation of a wider populace like students and other civilians, momentum needed to bring about required changes may remain elusive no matter how much focus is placed upon subsequent tracking endeavors.

Monitoring progress towards achieving goals set by communities regarding their response to global warming in mobilization of civil society and youth climate activism is essential for successful implementation (Chu, 2018; Cody, 2018; de Macedo & Jacobi, 2019; Fitch-Roy et al., 2020; Orach & Schluter, 2016; Zhan, 2022). Studies have shown that large-scale mobilization of citizens and the organization of grassroots campaigns are key components for citizen engagement in addressing climate change (Gardner et al., 2014). Additionally, research suggests that effective mobilization requires an understanding of how current systems work as well as a recognition or acknowledgment from those at policy level (Lahsen & Brondizio, 2008). One example is Mexico's General Law on Climate Change which includes input from over 500 civil society organizations including NGOs, academics, representatives from indigenous peoples' organizations and social networks (Reyes et al., 2011). This process has been used to build public support around initiatives such as the 'Youth For Climate Justice Campaign' which focuses on educating young people through workshops about environmental protection issues related to global warming (Moyo & Gyimah-Boadi 2016). In contrast to traditional state centric approaches where governments lead efforts alone; this type of networked governance creates new opportunities for collaboration between stakeholders with different interests resulting in more comprehensive solutions than what states can provide independently (Chapin III et al., 2013). Networked Governance opens up channels for communication among actors leading to better informed decisions across diverse groups while enabling greater participation in decision making processes. By engaging a variety of stakeholders it also allows individuals within each group who may not be part of formal governance structures but still hold local knowledge or expertise relevant to solving particular problems associated with climate change – such as farmers affected by extreme weather events – contributing meaningfully into discussions providing valuable insights. In effect then this kind of collaborative action offers powerful potential when combined with monitoring tools capable of identifying targets achieved along with areas needing improvement helping direct much needed resources quickly and efficiently reflecting changes required at both national and international levels.

CHAPTER 3 SUMMARY

In conclusion, the mobilization of civil society and youth climate activism has played a pivotal role in the global movement toward addressing the threat of climate change. Through awareness-raising initiatives, targeted campaigns, and creative strategies, these activists have effectively captured the attention of the public, media, and policy-makers, thereby significantly shaping the overall discourse on climate change. The undeniable impact of these groups can be attributed to their collective strength, passion, and commitment to a common cause, which exemplifies the power of grassroots activism to drive positive change within any society. As we transition from examining the forces that propel the climate movement to assessing the frameworks in place to respond to this global challenge, it becomes evident that collaborative efforts are vital for meaningful progress. The next chapter, "Networked Climate Governance," delves into the intricacies of how various actors – ranging from governments to international organizations to civil society – interconnect and collaborate in order to address the multifaceted nature of the climate crisis. By exploring the ways in which these networks function and succeed or fail in their endeavors, a deeper understanding of the practical aspects of climate governance can be obtained. With this knowledge, we may be better equipped to identify the areas in need of improvement, ultimately contributing to the development of more robust and effective climate policies and governance frameworks.

CHAPTER 4

Networked Climate Governance

CONTENTS

In an era marked by the exponential increase in environmental concerns and unprecedented global interdependence, the need for an effective yet inclusive approach to climate governance is imperative (Bowen et al., 2017; Brandenberger et al., 2022; Brewer, 2013; Brewer et al., 2017; Buchtmann et al., 2022). Traditional, centralized models of governance are increasingly under strain as they struggle to accommodate diverse actors, complex interactions, and transboundary issues. This has given rise to the emergence and growing relevance of "Networked Climate Governance," a concept that acknowledges the crucial interplay between various public and private actors in pursuing collective climate action. This chapter delves into the intricacies of networked climate governance, exploring its potential in shaping climate policies and strategies in alignment with global sustainability goals. Historically, climate governance was driven and managed by supranational entities, primarily intergovernmental organizations and national governments, through top-down regulatory frameworks. However, due to the decentralized nature of the problem and the multiplicity of actors involved, conventional governance approaches often fall short in addressing issues of global scale like climate change. This gap is further exacerbated by ever-evolving geopolitical dynamics, which can hinder international cooperation and render reactive climate policies inadequate. The concept of networked climate governance offers a fresh perspective by emphasizing inclusive, multi-layered collaboration among diverse stakeholders. These include national, subnational, and local governments, supranational organizations, private enterprises, non-governmental organizations, and civil society actors. By fostering partnerships and promoting collective action, this approach aims to create synergies between various policy instruments and facilitate realignment of global climate objectives. Ultimately, nested governance models are seen as important stepping stones in achieving more complex, transnational governance arrangements and upholding environmental integrity.

This chapter will walk the reader through the mechanisms and structures that underpin networked climate governance. It will first shed light on the rationale behind this paradigm shift, underlining the limitations of conventional governance models and their unsuitability to address the 21st-century climate challenges. Next, the chapter presents an analysis of key principles and characteristics that define networked governance, such as inclusion, adaptability, and polycentricity. These principles, as the chapter argues, make networked governance a more agile and innovative way of addressing climate change. Moreover, the chapter will delve into the dynamics of climate governance networks themselves, exploring different modes of stakeholder engagement, their interdependencies, and how the degree of convergence or divergence between goals, interests, and stakeholder capacities affect the outcome of collective climate action. Real-world examples will be provided to highlight the nuances of networked climate governance and its potential in facilitating the design and implementation of effective policies and strategies. Lastly, while recognizing the transformative potential of networked climate governance in addressing global climate challenges, this chapter acknowledges its pitfalls and caveats. The analysis will also identify the obstacles that these networks may face, such as power imbalances, expertise disparities, and inefficiencies caused by overlapping memberships, which could hinder the ultimate goal of safeguarding the environment. The discussion will culminate in providing recommendations and policy implications to enhance the effectiveness of networked climate governance towards a sustainable future. In sum, this chapter invites the reader to consider the pivotal role of networked climate governance as an essential cog in the machine of collective climate action. Through detailed analysis of its principles, real-world applications, and potential pitfalls, this chapter guides the reader towards a deeper understanding of networked governance and its capacity to address the pressing issue of climate change. Ultimately, the discourse aims to foster open-mindedness towards adopting multifaceted, collaborative approaches that strengthen the global response to the climate crisis.

4.1. COLLABORATIVE APPROACHES TO GOVERNANCE

The five principal aspects of grassroots community organizing in collective climate action and networked climate governance are as follows (Ensor & Harvey, 2015; Henry & Vollan, 2014; Kellner, 2023; National Intelligence Council, 2017): A central dimension in the discourse on the subject of collective climate action and networked climate governance refers to grassroots community organizing, which presents itself as an indispensable driving force behind these two movements. Using the case study of the Transition Towns movement inaugurated in Totnes, Britain, by Rob Hopkins and Naresh Giangrande in 2006, we can illuminate the principal aspects of such grassroots organizing in collective climate action. Transition Towns is a globally-networked initiative aimed at equipping communities to address climate change and peak oil through innovation at the local level, in doing so embracing a ground-up, collective model. The essence of this model lies in community participation that fosters organic, locally-rooted projects which pursue resilience and sustainability.

A critical component of the Transition Town model is the 'Reskilling' process, allowing communities to regain knowledge and abilities considered as 'old-fashioned' but are in fact crucial for non-fossil fuel reliant lives, such as food preserving and gardening. This community-driven knowledge sharing is a testament to a decentralized form of climate governance. Concurrently, thanks to the internet and other digital technologies, this local knowledge is shared globally, leveraging networked climate governance on a wider scale than ever before and thus, creating a roadmap for other communities aspiring to make a similar eco-transformation. By closely engaging with stakeholders, particularly local governments, the Transition Towns movement illustrates a democratizing influence on climate governance, proving grassroots community organizing an essential element in the grand scheme of global climate action.

Networked climate governance is a collaborative approach to addressing global warming. It utilizes the power of networks and collaboration among multiple stakeholders, including governments, civil society organizations, international bodies, businesses, academics and research institutions (Grimm & Schauffler 2016). This type of governance has been found to be effective at finding innovative solutions to complex environmental challenges such as reducing greenhouse gas emissions and promoting renewable energy sources (Seyfang et al., 2017; Howell et al., 2013).

In addition, networked approaches can also provide increased access for marginalized stakeholders who may not have had traditional decision-making authority in more hierarchical structures (Klinsky & Cashore 2007). By expanding the scope of participation in climate mitigation efforts beyond just government actors, this structure allows for greater collective knowledge sharing which can lead to more comprehensive policy development that considers local realities. This collaborative form of governing has become increasingly popular due to its ability to create multi-level partnerships between different groups while still allowing individual agency over decisions related directly to their own contexts or interests. Co-governance takes these same principles one step further by creating an even stronger link between all involved parties through shared responsibility for implementation and enforcement mechanisms (Shimeles 2019). All members are then actively engaged in making decisions about how best manage resources sustainably with mutual respect and understanding amongst each other's perspectives on what should be done towards achieving desired outcomes. This requires a systematic process for monitoring, reporting, and verification (Figure 4.1).

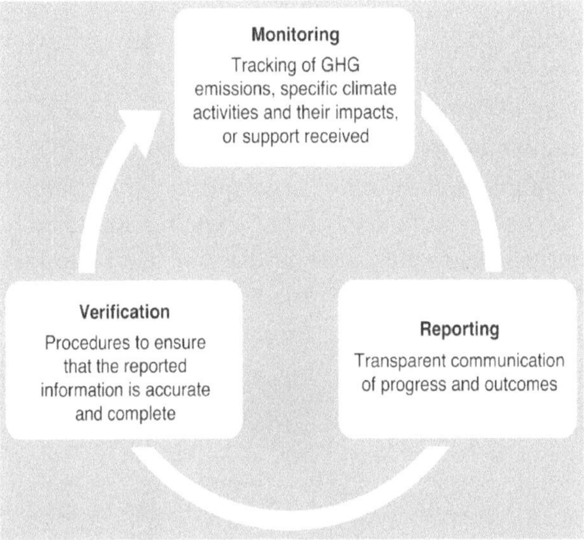

Figure 4.1. The cycle of reporting, verification, and monitoring in climate policy compliance.

Source: Cambridge University Press & Assessment.

Co-governance in networked climate governance is a term used to describe the collective action of multiple actors, working together to address complex environmental issues such as climate change (LeBel et al., 2016). Co-governance has been found to be effective when it is based on shared decision making between different levels and stakeholders within a particular geographical area or region. For example, Lebel et al. (2016) studied co-governance initiatives involving local governments and Indigenous communities in Canada's North where collaborative efforts yielded successful outcomes for conservation management of natural resources. In addition, research by Binder & Cashore (2018) suggests that adaptive governance approaches are highly compatible with those of co-governance as both emphasize collaboration among various stakeholders while applying an iterative approach towards problem solving. Moreover, according to their study, self-organizing networks have emerged from these collaboratively designed policies which are able to effectively respond quickly and flexibly against changing conditions due to higher levels of trust established through increased communication amongst members involved in the initiative. Adaptive governance refers specifically to an approach towards managing socio ecological systems which focuses on developing flexible responses over time using experimentation and learning from past mistakes (Bodin & Crona 2008). It involves continuous monitoring, feedback loops incorporating relevant stakeholder perspectives; whereby each actor can adjust goals accordingly depending upon ongoing changes within the system(s), allowing for greater flexibility compared with traditional top-down command/control methods adopted by policy makers historically (Gunderson et al., 2002). Adaptive Governance thus promotes collaboration across sectors leveraging existing knowledge bases along with data gathering techniques relating directly back into public policies at different scales enabling more informed decisions about how best manage our environment collectively going forward.

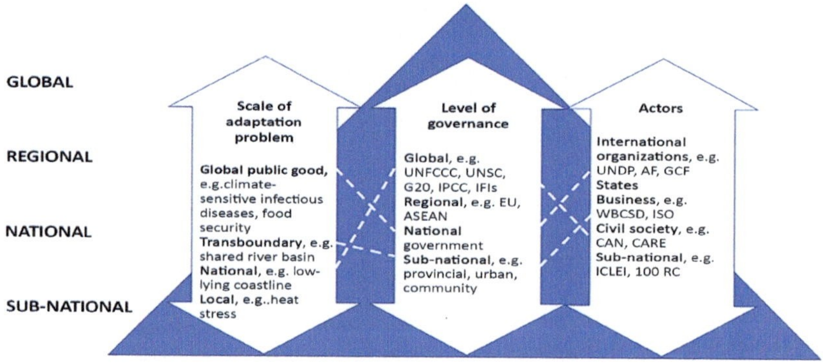

Figure 4.2. Climate governance structures interaction at various global levels.

Source: Wiley Interdisciplinary Reviews.

Networked climate governance is an approach to managing the environment that involves interaction between multiple actors, including industry, governments, civil society organizations and local communities (Figure 4.2). Adaptive governance is a form of networked climate governance which incorporates both existing policy tools (such as emissions trading schemes) and non-traditional approaches such as citizen science initiatives (Bennett et al., 2016). It recognizes the complexities of environmental decisions by emphasizing collaboration and learning over time. Through adaptive management strategies such as monitoring feedback loops, it allows for flexible responses to changing conditions in order to achieve desired outcomes more effectively (Folke et al., 2005). Adaptive governance has been increasingly used in recent years due to its potential effectiveness at addressing complex problems like global warming. For example, researchers have found that when using this approach there can be improvements in ecosystem health (Kauffman & Stevens 2010), better understanding of how different stakeholders interact with each other on issues related to land use changes (Brown et al., 2017), improved resource efficiency through collective action among farmers adapting their practices based on shared knowledge about weather patterns (Tompkins & Adger 2004), or reduced risk exposure from extreme weather events caused by climate change (Munroe 2015). Additionally, studies show that countries implementing adaptive policies have seen significant economic benefits compared with those who do not implement them – for instance in terms of greater gross domestic product growth rate per capita during times of crisis resulting from natural disasters induced by climate change shocks (Galiani et al., 2020).

Overall then, adaptive governance can provide effective solutions for dealing with complex challenges posed by global warming (Partelow et al., 2020; Pavanelli et al., 2023; Perz et al., 2022). By allowing stakeholders involved in decision making processes around these issues work together collaboratively while also being open to new ideas and experimentation; they are able create dynamic systems capable responding quickly adaptively respond appropriately given current conditions On top of this, adapting old methods like emissions trading schemes ensures traditional forms regulation still play role alongside newer techniques focus on stakeholder engagement education exchange information further increases chances success tackling major problem facing humanity today: Climate Change. Networked Governance refers generally to a system organization where many parties cooperate voluntarily coordinate activities resulting in a common goal without need of central authority control structure affecting participation decisions taken within group centralized body state. In context, the Network Climate Governance concept applies specifically to interventions strategies implemented to mitigate effects of human activity leading to increased atmospheric temperatures causing drastic shifts in Earth's ecosystems. This requires coordination efforts of various public private sector organizations to address issues at all levels of regional and international scales.

Climate change is a complex issue that requires collaboration between multiple stakeholders from different countries and regions (Bernstein & Hoffmann, 2018; Bixler et al., 2023; Bjerkan & Ryghaug, 2021; Blok, 2018; Bodin, 2017; Bodin et al., 2019). As such, transnational governance networks (TGNs) have emerged as an important tool for addressing global climate issues through collective action. TGNs are multi-stakeholder initiatives which bring together actors from government, civil society, business sector, academia and other organizations to work on solutions to global challenges like climate change (de Coninck et al., 2017). Such networks provide the necessary platform for dialog among these actors in order to develop joint strategies or policy responses.

They also increase access to resources and facilitate knowledge sharing across borders (Lemos & Moreira 2007). These benefits of TGN's make them effective agents in networked climate governance – providing both collaborative problem solving capacity within individual nations as well as facilitating sustainable international cooperation regarding long term environmental protection goals (Gormley & Lane 2012). At the same time, it is important to note that while Transnational Governance Networks can be powerful tools for tackling global problems such as climate change they are not always sufficient by themselves; other forms of Multi-level Governance may need to be employed alongside them if maximum progress is going to be achieved. Multi-level Governance involves coordination among nation states at multiple scales including regional level agreements or even supranational institutions like the European Union (EU) with its various policies regulating carbon emissions etc. (Scholten et al., 2016). On their own however TGN's offer an invaluable resource pooling opportunity allowing high levels of participation from all relevant parties whether national governments or grassroots activists thus making them key players in Networked Climate Governance efforts worldwide.

4.2. MULTI-LEVEL GOVERNANCE

Climate change, as a global issue of immense complexity, presents a formidable challenge that transcends the conventional understanding of governance (Butler et al., 2022; Hawkins et al., 2016; McLaughlin, 2022; Partelow et al., 2020; Sultana et al., 2019). The interconnected nature of the drivers, impacts, and solutions demands innovative approaches to addressing it. One such approach that has gained prominence in recent years is "Networked Climate Governance," which emphasizes the importance of collaboration among diverse public and private actors for effective climate action. Particularly, the concept of multi-level governance plays a critical role in networked climate governance, proposing a framework for coordination and cooperation across different hierarchical levels. This section will lay the foundation for understanding multi-level governance within the context of networked climate governance. By examining the theoretical underpinnings and key principles of this approach, we will reveal its potential and challenges in fostering collective action and achieving climate goals. Multi-level governance, as a concept, emerged primarily from the recognition that traditional top-down, centralized models of governance often struggle to address intricate and transboundary issues such as climate change. This failure has given rise to the development of more inclusive, horizontal, and vertical governance arrangements that involve a multitude of actors at various levels, ranging from international organizations and national governments to local authorities, non-governmental organizations, and private enterprises. By encompassing these diverse stakeholders, multi-level governance aims to foster collaborative decision-making processes and enhance the responsiveness, flexibility, and effectiveness of climate policies.

We will investigate the critical elements that characterize multi-level governance, such as decentralization of decision-making, subsidiarity, and polycentricity. This investigation will highlight the relationships and interdependencies among various actors and how these interactions facilitate dynamic policy-making and implementation processes. Furthermore, we will outline several examples of multi-level governance in networked climate governance, demonstrating how this approach has been adopted and operationalized within diverse contexts and policy domains. Despite the potential of multi-level governance, it is essential to acknowledge the challenges that may arise in its application. We will examine the potential obstacles to effective multi-level governance, including coordination difficulties, power imbalances, and resource constraints, all of which may hinder collective action and policy implementation.

Exploring these challenges will help us to understand the potential limitations and areas for improvement in multi-level governance approaches. Finally, we will consider the implications of multi-level governance for networked climate governance, reflecting on how this approach contributes to overall climate governance goals. By exploring the interactions between multi-level governance and other governance initiatives, we will draw connections and identify synergies that can propel collective climate action (Figure 4.3). In conclusion, this section will establish the foundations for examining multi-level governance in networked climate governance. By unraveling the key principles, opportunities, and challenges presented by this approach, we will deepen our understanding of the potential of multi-level governance in addressing the complex and pressing issue of climate change. Ultimately, our exploration will provide valuable insights into how multi-level governance can contribute to the development of more effective, inclusive, and coordinated climate action.

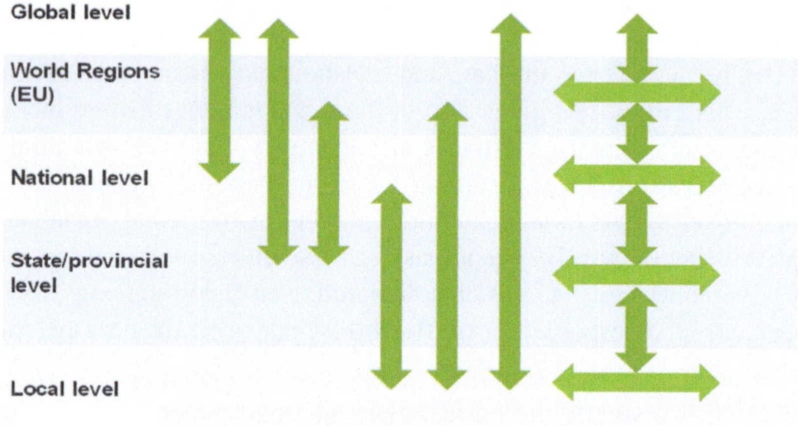

Figure 4.3. The multiplicity of possible interactions in the model of global multi-level governance.

Source: Jänicke (2017).

The concept of multi-level governance (MLG) is an important tool in analyzing how different levels of organization interact with one another to achieve shared goals. It has been used to explain the emergence and conceptualization of a wide variety of complex social systems, including those related to health care policy, urban development, financial regulation and environmental management (Lowndes & Pratchett, 2008; Nalepa & Bressers, 2010). MLG involves understanding both horizontal linkages between different actors within each level as well as vertical relationships between higher and lower tiers or regions (Söderbaum et al., 2011). These interactions are informed by power dynamics that can be seen at all scales – from international relations between countries down to local negotiations among stakeholders over specific projects (Kooiman et al., 2009). Emergence occurs when these multiple layers come together in unplanned but coordinated ways, creating something greater than the sum of their parts. Through this process new patterns emerge which may not have previously existed before (Teece et al., 1997). This organizational complexity presents challenges for analysis since it often produces outcomes that cannot be predicted based on any single actor's behavior or decisions alone. Thus, emerging solutions must also take into account existing structures and norms as well as other factors such as economic conditions or cultural values (Brenner & Theodore 2002; Gaventa 2006). By looking beyond individual actions towards broader systemic changes we gain insight into how change can happen across multiple scales simultaneously.

Multi-level governance provides a useful framework for exploring the emergence and conceptualization of complex social systems because it allows us to consider issues holistically while still accounting for differences in scale and scope (Davies et al., 2018; Deka & Goswami, 2022; Dey et al., 2017; Heikkila & Gerlak, 2019; Tosun & Schoenefeld, 2017). Understanding MLG requires considering key principles such as subsidiarity – whereby decision making should occur at the lowest level possible -and proportionality – where decision makers take into account potential impacts on various groups affected by their choices (Tinbergen Institute 2015). In addition, characteristics like transparency, accountability, participation, inclusion, coordination amongst others play an essential role in determining success. The ability to identify connections among components at various levels helps us create meaningful interventions that address underlying causes rather than just symptoms. Ultimately then thoughtfully applying concepts associated with MLG leads us closer towards achieving sustainable progress both locally and globally.

Multi-level governance (MLG) is a concept that has grown in popularity over the past few decades, and it describes how different levels of government interact with one another to create better policy outcomes. MLG is based on three key principles: subsidiarity, interdependence, and multilayered decision-making (Scholz & Kübler 2018). Subsidiarity refers to the idea that decisions should be taken at the lowest possible level by those closest to them; this ensures more efficient problem solving as local actors have greater knowledge of their specific context than central governments do. Interdependence recognizes that different layers of government often rely upon each other for resources or support in order to effectively implement policies; this encourages collaboration between various levels of government rather than competition amongst them. Lastly, multilayered decision making acknowledges that multiple organizations will need input into any given policy issue in order for consensus and compromise solutions to emerge (Köllner et al., 2013). Real world applications and examples can be seen all around us when we examine issues such as healthcare reform or educational funding initiatives. In terms of healthcare reform, there needs to be cooperation between state legislatures who are responsible for setting up insurance exchanges along with federal agencies like Medicaid who provide additional subsidies so people can access care more easily (Sommers et al., 2016). Similarly, when looking at education spending initiatives from both national and local governments, multi-level governance allows for increased flexibility across states while also maintaining compliance standards set out by higher authorities through collaborative efforts among school boards administrators and governors offices (Marmor 2017). These types of synergistic relationships demonstrate how important it is today for our society's most pressing problems – like health care or education -to leverage multi-level governance strategies if successful public policies are going to be crafted (Figure 4.4).

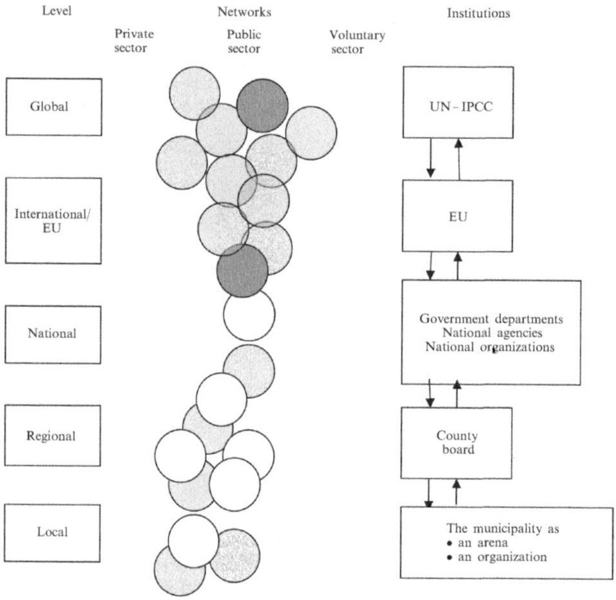

Figure 4.4. Multilevel governance and climate change policy.

Source: Gustavsson et al. (2009).

Multi-level governance (MLG) is a concept that has revolutionized the way governments and other stakeholders interact in international politics, policy formulation and implementation. MLG involves multiple actors at different levels of government – local, regional, national and supranational institutions – working together to achieve common goals (Stoker & Skelcher, 2020). It is seen as an effective mechanism for advancing global objectives such as poverty reduction or sustainable development since it allows for more coordination between public authorities across boundaries while allowing each jurisdiction's autonomy on matters within its own competence (Folke et al., 2005). However, this model presents several challenges. Firstly there can be difficulties implementing decisions due to differences in legal regulations between jurisdictions which complicate enforcement efforts by creating gaps where laws are enforced differently from one place to another (Haufler & Lake, 1999). This lack of consistency makes it hard to ensure compliance with any agreements made by all parties involved. Moreover, diverging interests among various stakeholder groups create obstacles when trying to reach consensus on issues that affect them (Agranoff & McGuire 2003). Furthermore, a key challenge lies in finding ways of involving civil society organizations into decision-making processes so they can provide alternative perspectives and represent citizens' concerns effectively without undermining state authority or weakening accountability mechanisms (Kooiman et al., 2007). Finally limited resources pose additional constraints; developing countries may struggle financially if expected contributions exceed their capacity or local communities fail to receive the support needed from higher level governments because fiscal transfers do not always flow down adequately (Lebel 2006). These limitations reduce effectiveness even further making it difficult for MLG initiatives to produce tangible results. Despite these drawbacks however, multi-level governance offers some advantages too; enabling collaboration between different levels of government encourages better use of available resources through economies of scale thus reducing administrative costs while also improving service delivery overall (Söderbaum et al., 2002). Additionally, cooperation facilitates access to previously unavailable knowledge sources thereby enhancing problem-solving capacities considerably (Van Bommel 2008). By taking advantage of synergies like these MLG initiatives have potential for overcoming many existing barriers towards achieving socio-ecological resilience despite facing inherent challenges along the way.

Multi-level governance (MLG) has become an increasingly popular framework for managing public policy and decision making in the 21st century. MLG is defined by Heikkila et al. (2006, p.3) as "the set of interlinked networks of formal and informal actors at multiple scales that are able to make decisions on a wide range of issues" which can span from local communities to national governments, international organizations and beyond. This approach offers many advantages over more traditional top-down models including greater flexibility in responding to changing conditions, increased efficiency due to better coordination between stakeholders, and improved accountability through multi-directional feedback loops (Heikkila et al., 2006). By leveraging these synergies among different levels of government it is possible to develop innovative solutions tailored specifically for regional contexts while still achieving nationally mandated goals (Scepanovic & Bojanic 2012). One area where MLG has seen particularly strong success is the development of public-private partnerships (PPPs). In recent years there has been a growing recognition that private sector involvement holds considerable potential when it comes to designing effective policies, delivering services more efficiently or mitigating risk associated with large scale investments (Molle & Meijerink 2011; Schouteten et al. 2015). Multi-level governance structures provide the perfect platform for facilitating this kind of collaboration as they enable both sides to bring their distinct strengths together under one roof thereby creating valuable opportunities for mutual learning and shared problem solving (OECD 2017). For example, according to Ebrahimnejad et al. (2015), local water management initiatives implemented within an integrated MLG structure were able to achieve significantly better outcomes than those relying solely upon centralized state funding by combining central resources with experience from community members who had direct knowledge about local needs. Thus we see how powerful synergistic effects can be unleashed when diverse groups work together towards common objectives facilitated through appropriate frameworks such as MLGs.

4.3. PUBLIC-PRIVATE PARTNERSHIPS

Tackling the multifaceted issue of climate change requires a collaborative approach that acknowledges the interconnected nature of global environmental challenges and the diverse array of stakeholders involved in addressing them (Buchtmann et al., 2022; Butler et al., 2022; Calliari et al., 2019). One of the key components of networked climate governance is the facilitation of cooperation among these stakeholders, particularly through PPPs. This section sets the stage for understanding how PPPs function within the context of networked climate governance and explores their potential in contributing to more effective and inclusive climate strategies. PPPs represent a collaborative governance model that combines the resources, expertise, and strengths of public and private sector actors in the joint pursuit of common goals, such as climate mitigation and adaptation efforts. By fostering cooperation between these different entities, PPPs have the potential to enhance the scalability, reach, and impact of climate initiatives—ultimately working towards more sustainable policy outcomes.

In this section, we will outline the key characteristics and principles of PPPs, highlighting their unique potential alongside traditional top-down governance models. We will explore how PPPs provide opportunities for innovative financing and resource allocation, knowledge and technology exchange, capacity building, and risk sharing, among other benefits. By understanding the dynamics of these partnerships, we will offer valuable insight into their role in networked climate governance. Furthermore, we will examine real-world examples of successful PPPs within the realm of climate governance. These case studies will illustrate how diverse stakeholders have come together in various contexts to jointly address climate change and related environmental challenges. By analyzing these collaborations, we will provide a better understanding of the conditions in which PPPs thrive and the potential barriers they may encounter.

	Government-Private company	**Government-Nonprofit**	**Government-private company-Nonprofit**
Project/Activity	Hospital project, School project	Public sector employees 'opt out' to form own non-profit project	Gate21 in DK, a partnership on climate change issues
Organizational form	PFI organizational form	Red Cross running refugee camps on behalf of government	Networked partnership across multiple organizations
Policy	UK Treasury PFI policy Canadian PPP policy Victoria, Australia PPP policy European Commission PPP policy across directorates from 2010-	'Big Society' agenda in UK	World Bank policy in private sector involvement
Governance tool or symbol	EU on the role of government in the European economy	Role of non-profits in delivery of specific services in the US because of pressing policy issues	Millenium goals in the UN
Historical context of public-private relations	Can market economy continue the same way after the financial crisis?	Associations' integral use in the building of the administrative state	UN Global compact between developed and developing world

Figure 4.5. Examples of forms of partnership (public-private) and dimensions of the PPP phenomenon.

Source: Hodge (2011).

Though PPPs offer a range of advantages in networked climate governance, it is essential to recognize their limitations and risks (Bodin et al., 2019; Brewer, 2013; Herzog & Ingold, 2019; Homobono et al., 2022; Pavanelli et al., 2023). Potential challenges such as power imbalances, the risk of regulatory capture, and the need for transparency and accountability within PPPs will be discussed, providing a more comprehensive perspective on the feasibility and sustainability of these collaborations. Finally, considering the broader implications of PPPs for networked climate governance, we will examine how PPPs contribute to advancing climate goals and supporting inclusive, coherent, and effective policy frameworks. By exploring the synergies and complementarities between PPPs and other governance mechanisms, we will provide insights on how these collaborations can become even more impactful components of networked climate governance. In conclusion, this section will establish the foundation for a comprehensive understanding of PPPs in the context of networked climate governance. By delving into the key principles, real-world applications, and potential challenges of PPPs, we aim to provide a balanced and neutral perspective on their role—and prepare the reader for a stimulating and informative exploration of this important component of climate governance.

Five key aspects emerge when analyzing PPPs in networked climate governance (Bailey et al., 2012; Brandenberger et al., 2022; Lee & Koski, 2014; Pavanelli et al., 2023; Rahimi-Feyzabad et al., 2022):

1. *Development and rationale*: The first principal aspect of PPPs in networked climate governance is their development and rationale (Figure 4.5). As limitations of traditional top-down governance models became evident in addressing complex global issues such as climate change, the need for more collaborative arrangements arose. PPPs emerged as a solution, combining the strengths of public agencies and private sector entities to create synergies and leverage resources more effectively in the pursuit of climate goals.
2. *Key principles and benefits*: The second principal aspect highlights the key principles and benefits of PPPs in networked climate governance, such as innovative financing, resource allocation, knowledge and technology exchange, capacity building, and risk sharing. By functioning on the foundation of these principles, PPPs can foster cooperative decision-making processes and enhance the overall effectiveness of collective climate action.
3. *Real-world applications and case studies*: The third principal aspect involves examining real-world examples and case studies to understand how PPPs operate within the realm of climate governance. By analyzing these collaborations across diverse contexts, insights can be gleaned on the conditions that favor the success of PPPs and the shared value they can create through their efforts.
4. *Challenges and risks*: The fourth principal aspect recognizes the potential challenges and risks associated with PPPs in networked climate governance. Power imbalances, risks of regulatory capture, and the need for transparency and accountability are some of the potential drawbacks of such collaborations. Acknowledging

these challenges provides a more comprehensive understanding of PPPs and their implications in climate governance.

5. *Implications and synergies*: The fifth principal aspect examines the broader implications of PPPs for networked climate governance and collective climate action. By exploring the connections, synergies, and complementarities between PPPs and other governance mechanisms, insights will be revealed about the potential of PPPs in propelling climate action, and how such partnerships can be more effectively integrated and deployed within the broader context of networked climate governance.

Multi-level governance (MLG) is a system of government in which multiple layers or units interact to manage public policy issues and concerns. It has emerged as an alternative to traditional centralized and monolithic modes of governing, offering increased flexibility, diversity, efficiency, responsiveness and accountability (Lodge & Wegrich 2018). MLG can be seen as having three key principles: the separation of roles between different levels of government; decentralization from central authorities; and coordination among these actors at various scales (Kerwer 2020). These principles are not hardwired but rather shaped by historical context such as particular political conditions or economic needs. Research suggests there are several benefits associated with multi-level governance systems. For example, Finotelli et al., (2018) found that regional governments offered greater capacity for responding quickly to local situations than did their national counterparts due to their proximity to citizens' everyday lives. Additionally, decentralized decision making on certain matters allows for more tailored policies better suited for specific contexts instead of one size fits all approaches often used under top-down models (Tavares et al., 2017). Furthermore, shared responsibilities among different levels allow them each to focus on what they do best while avoiding duplication leading to improved efficiency overall (Haque & Nadvi 2015). Finally, studies have shown that MLG expands opportunities for citizen participation resulting in increased social trust and legitimacy towards government institutions (Nicolau 2016).

Multi-level governance (MLG) has become an increasingly prevalent approach to the design and implementation of public policies in recent decades (Hailbronner & Koppenjan, 2017). MLG seeks to better coordinate resources among different governmental entities through shared decision making while also providing for more flexibility within each level's autonomy (Loughlin et al., 2013; Scholten et al., 2014). This type of governance promotes innovation by allowing all parties involved to have input on decisions related to their area of expertise or interest. In addition, it offers enhanced accountability because stakeholders at various levels are responsible for ensuring that objectives are met as well as creating incentives for collaboration between jurisdictions with divergent interests (Fung & Wright 2003). These key principles provide many potential benefits when applied in real world settings. For example: MLG can help reduce overlap or duplication in services provided by higher level governments which can lead to cost savings overall (Scholten et al., 2014); facilitate greater coordination between local initiatives such as urban planning projects and state/national development goals (Figure 4.6 for the principles underpinning this); create clearer lines of responsibility regarding who is accountable for what types of outcomes; improve public participation since citizens may be able to get closer access to decision makers whose actions affect them directly; and finally promote effective problem solving through collective action where solutions created together often yield stronger results than those achieved separately due only unilateral efforts from one entity (Vogel 2007). Real world applications include cases like the European Union's multi-level system which allows member states some degree of control over issues concerning their own constituencies while simultaneously maintaining unified regulations throughout its common market (Majone 1996). Another example would be how Canada incorporates both Provincial Governments alongside Indigenous First Nations groups into crafting environmental protection policies intended to protect fragile ecosystems from human activities such as oil extraction operations or logging practices (Boffey 2019). These two examples demonstrate the power behind MLGs ability not just foster cooperation but also ensure desired outcomes are reached without sacrificing individual rights along the way.

Number	Partnership Principle
1	Purpose is clear, aligned and realistic
2	Availability of appropriate financial and human resources
3	Clarity of motivations, roles, capabilities and contributions
4	Sufficient organisational processes and procedures that foster collaboration
5	Alignment of partners and policies
6	Commitment, ownership and responsibility of partners towards the partnership
7	Partnership is participative and empowering
8	Culture of collaboration trust and openness
9	Presence (and awareness) of cultural transformation, synergy, efficiencies or exchange
10	Defines success monitors and reports its performance
11	Partnership is continually engaging with others, developing and learning
12	Clear attribution of benefits, risks and blame

Figure 4.6. Key principles of public-private partnerships (PPPs).

Source: Jeffares et al. (2009) via Hodge (2011).

Multi-level governance is a system of governing that combines both national and local levels, with the objective to provide better public service outcomes for citizens (Ostrom & Ostrom, 2008). Real-world applications have seen multi-level governance implemented in cities around Europe. For example, Kooiman et al. (2014) found evidence of successful multi-level governance working together between regional governments and municipalities in Germany through shared knowledge and joint initiatives such as sustainability programs. Similarly, Olshansky (2012) studied the effects of multi-level governance on decision making processes within Scotland's devolved government structures which saw increased collaboration between different tiers of government leading to improved policy outcomes including laws relating to renewable energy sources. The empirical data also shows clear benefits from applying Multi level Governance approaches; according to Huberts et al., 2019), implementation led to an increase in community involvement resulting in higher efficiency due largely to more comprehensive consultations held across all involved parties throughout various stages of project development life cycles. The advantages associated with Multi Level Governance are undeniable. However, it does present certain challenges and risks when applied without proper consideration or planning given its complexity (Huberts et al., 2019). In particular, there can be difficulties managing conflicts amongst stakeholders at multiple levels if not handled correctly by those responsible for overseeing these projects along with issues related to ensuring sufficient resources are allocated appropriately during each phase especially considering stretched budgets often encountered within modern public sector settings (Kooiman et al., 2014). Therefore, careful management should always be taken into account prior implementing any form of multilevel governance approach otherwise this could lead to potential failure before it even begins thus undermining the entire process altogether potentially taking away some much needed positive changes being made towards our society today.

Multi-level governance (MLG) is a type of decision-making structure that includes multiple actors from different levels and locations in policy formation, implementation, and evaluation (Fidelis et al., 2019; Lee et al., 2014; Lubell & Robbins, 2022; Orach & Schluter, 2016; Pavanelli et al., 2023). MLG exists at international, regional, national, subnational and local levels depending on the context. It is often used to address complex issues like climate change or global poverty which require cooperative efforts between various stakeholders (Hoffmann et al., 2019). As such it can provide more effective solutions compared with traditional hierarchical structures or single level approaches due to its ability to tap into diverse resources across many jurisdictions (Saward & Bevir 2011). However, there are also challenges associated with multi-level governance as well as risks for those involved in implementing them.

One significant challenge faced by MLG systems is coordination amongst actors from different tiers who have varying interests and priorities (Börzel & Risse 2012; Dijkstra 2017). This means that despite having access to greater expertise via cooperation between higher level authorities such as states or provinces down through lower tier bodies including cities or townships – differences in opinions may create tension when reconciling these divergent viewpoints during negotiations leading up to decisions being made (Rocco 2020). Moreover trust must be established between all parties before progress can be made considering the need for mutual understanding required within participatory processes along each step of the way involving numerous players both inside and outside government agencies (Cairney 2018; Scholten 2015). Ultimately this makes reaching consensus difficult since participants will always try to pursue their own particular agendas regardless if they support collective objectives effectively hindering achievement of common goals set out initially (Scharpf 1999; Sabel 2008). Furthermore, unclear roles amongst partners coupled with unbalanced power dynamics further complicate matters making compromise harder still while creating potential conflicts over jurisdiction rights regarding responsibility for outcomes obtained by action taken (Eisele 2014; Strang 2010). Without proper management of expectations combined with sufficient monitoring risk factors become prominent thereby increasing chances of failure especially given limited capacity found among lower tier actors responsible for implementation.

Despite these difficulties however synergy effects generated through collaboration between different layers offer opportunities which should not be neglected either. For instance, sharing information allows all sides involved to gain insights that are useful towards forming better policies based on sound research rather than speculation resulting in improved effectiveness overall (Koskenniemi 2002; Leibfried 2009). Additionally pooling resources provides an opportunity to increase efficiency allowing funds allocated dedicated specific tasks thus preventing any one group going over budget while avoiding duplication elsewhere too saving time money effort expended unnecessarily otherwise (Loughlin 2005; Stoker 2006). In conclusion, although MLGs present certain obstacles related to coordinating large numbers dynamic entities plus managing risk inherent therein synergies created working together outweigh drawbacks ultimately reaping rewards yields benefits far exceed those achieved acting alone.

Multi-level governance (MLG) is an approach to public policy that recognizes the interconnectedness of political, economic and social interests at multiple levels. MLG seeks to achieve synergy between different layers of government through a system of cooperation rather than competition (Lorenz & Wollmann, 2009). This means that policies are developed in collaboration with local governments as well as state and federal agencies, allowing for more effective coordination among different administrative entities. In addition, it allows for greater citizen participation since decisions are not made solely by central authorities but also involve representatives from various communities and constituencies throughout the country. A key outcome of multi-level governance has been improved efficiency in decision making processes due to increased transparency and accountability (Krekula et al., 2011). For example, research shows that countries with decentralized political systems have higher rates of GDP growth compared to those with centralized systems (Eggerickx & Lefebvre de Buncey 2008; Voors et al., 2012). Moreover, studies indicate that decentralization can lead to better service delivery outcomes such as lower corruption rates and reduced poverty levels (Shenoni 2014; Corrales 2015).

Therefore, MLG plays an important role in promoting sustainable development objectives by facilitating efficient resource allocation across sectors while improving access to services for citizens living in rural or remote areas. The concept of circular economy is closely linked to multi-level governance because its success depends heavily on how effectively regulatory measures are implemented at all relevant scales. The idea behind this model is that resources should be reused instead of discarded whenever possible so they remain within productive cycles without ending up as waste products (Stahel 2020). Thus far however implementation efforts have focused primarily on production level activities which limit our ability to capture systemic synergies across entire value chains including consumer behavior patterns at end user stages (Rydin 2019). To fully realize these potential benefits it will be necessary for governments around the world to adopt integrated approaches involving both top down regulations alongside bottom up initiatives targeting changes in individual consumption habits over time. In conclusion then Multi-level Governance offers significant opportunities when applied strategically towards achieving sustainable development goals like those associated with Circular Economy models. This is notably achieved via improved decision-making and community involvement. Hence, the dynamics driving this approach shift away from linear forms of disposal centricity and toward regenerative solutions with participation-oriented pathways.

4.4. CIRCULAR ECONOMY AND SUSTAINABLE DEVELOPMENT

Amidst the growing global concern for environmental issues and the urgency to address climate change, sustainable development has emerged as a guiding principle in formulating multidimensional policy responses (Lee & Koski, 2014; Leventon et al., 2017; Lubell & Robbins, 2022; Luthe & Wyss, 2016). Within this context, the notion of a circular economy has gained significant attention as an innovative approach to promote resource efficiency, reduce waste, and foster environmental stewardship. In this introduction, we will explore the concept of the circular economy and its potential contribution to sustainable development in the context of networked climate governance, while maintaining a neutral tone. The circular economy represents a paradigm shift from the traditional linear economy (characterized by a "take, make, dispose" model) to a more regenerative and restorative economic system. At its core, the circular economy seeks to minimize resource consumption and waste generation by design, emphasizing closed-loop systems, extended product life cycles, and innovative business models that encourage reuse, recycling, and upcycling of materials. By reducing dependence on finite resources and mitigating environmental impacts associated with resource extraction, production, and disposal, the circular economy aims to contribute to both social and environmental goals of sustainable development.

Networked climate governance, as a collaborative and inclusive approach to addressing climate change, aligns well with the principles of a circular economy and sustainable development (Currarini et al., 2016; Dal Poz et al., 2022; Deka & Goswami, 2022; Gordon, 2016; Koontz et al., 2015). By fostering cooperation and collective action among various public and private actors, networked climate governance has the potential to facilitate the adoption and implementation of circular economy strategies across diverse sectors, regions, and stakeholder groups. Several aspects of networked climate governance can contribute to the promotion of a circular economy and sustainable development. For example, knowledge sharing and capacity building initiatives within networked governance arrangements can facilitate the dissemination of best practices and innovative solutions related to circular economy principles, thereby accelerating their adoption.

Furthermore, multi-stakeholder collaborations and PPPs can encourage the development of supportive policies, frameworks, and investment opportunities that can drive the transition to a circular economy. However, it is important to acknowledge the challenges and limitations that may arise in attempting to align networked climate governance with the goals of a circular economy and sustainable development. Potential obstacles may include regulatory inconsistencies, vested interests, trade-offs between economic development and environmental protection, and the need for technological advancements to enable circularity in various industries. A neutral appraisal of the circular economy must recognize these challenges and seek ways to address them within the context of networked climate governance. Overall, the circular economy and sustainable development represent crucial aspects of contemporary discourse on climate governance, and their alignment with networked climate governance presents opportunities for more efficient resource use and waste reduction. By fostering cross-sectoral collaboration and innovative partnerships, networked climate governance can play a pivotal role in advancing the transition to a circular economy in support of sustainable development goals. As we explore this intersection, it is essential to remain cognizant of both the potential benefits and the inherent challenges this integration might entail. Figure 4.7 is an example of the transition from a linear to circular economy in Serbia.

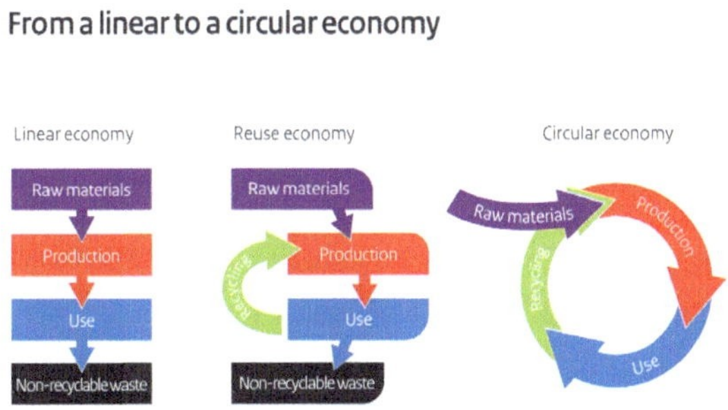

Figure 4.7. Transitioning from a linear to circular economy in Serbia.

Source: United Nations Development Program.

Five aspects of the circular economy and sustainable development emerge including (Willems & Giezen, 2022; Wyborn, 2015; Yates, 2012): The first principal aspect of Circular Economy and Sustainable Development in Networked Climate Governance is resource efficiency. This refers to the optimization of resources used, minimizing waste and energy consumption, and reducing greenhouse gas emissions. Achieving greater resource efficiency in a circular economy requires collaboration between diverse stakeholders, such as governments, businesses, and non-governmental organizations, to create and implement strategies for sustainable production and consumption. In recent years, the concept of circular economy has been gaining traction in sustainable development policy circles as a means to reduce environmental degradation and promote economic growth. According to the World Economic Forum (2020), "a circular economy is an industrial system that is restorative or regenerative by intention and design" which aims to keep resources within closed-loop systems for longer duration cycles. This approach seeks to minimize waste production through product reuse and recycling while also improving resource efficiency; thus reducing energy costs associated with traditional linear models of producing goods from raw materials. As part of collective climate action strategies towards global sustainability, initiatives such as networked climate governance are essential components for ensuring effective implementation of such approaches. A key element here lies in establishing shared responsibility among stakeholders including governments, businesses, civil society organizations – all working together to create systemic change on multiple levels (Klein & Schaefer 2019). The notion of conservation embodied in these concepts can easily be linked back to principles underlying waste reduction measures like recycling programs. By creating incentives for individuals and corporations alike to engage with more efficient production practices through technological advancements, we not only help preserve natural ecosystems but also stimulate economic activity at large scale (Grantham et al., 2018). While this type of collaborative effort will require ongoing investments from both public and private sectors into research/development activities related to eco-innovation – it stands out as one way forward towards achieving greater resilience across diverse socio-ecological contexts around the world today.

The second aspect involves reducing waste generation and promoting recycling and reuse of materials (Davies et al., 2018; de Macedo & Jacobi, 2019; Deka & Goswami, 2022). Networked climate governance plays a crucial role in creating platforms for exchanging best practices, policies, and technologies across various jurisdictions and sectors. The implementation and promotion of waste management systems, recycling facilities, and circular business models contribute not only to a closed-loop economy but also to achieving Sustainable Development Goals. In recent years, the concept of a circular economy has emerged as an important approach to reducing waste and increasing recycling. According to the European Commission (2018), "a circular economy is based on three core principles: design out waste and pollution; keep products and materials in use; regenerate natural systems." This focus on resource efficiency helps reduce negative environmental impacts while still allowing for economic growth. By optimizing resources through reuse, recycle or upcycling strategies, valuable raw materials can be recovered from existing technology instead of being extracted from nature (Klöpffer & Weber 2020). Additionally, by reusing inputs rather than relying solely on new ones we are able to conserve energy sources (Dellinger et al. 2018). The United Nations Sustainable Development Goals provide an essential framework for understanding how collective climate action plays a role in this process. For instance, Goal 12 calls for sustainable consumption and production patterns that seek to reduce global warming gasses at their source (UN 2017). Through implementing such sustainability practices across multiple sectors like energy management and manufacturing processes businesses can create both cost savings opportunities as well as help mitigate harmful emissions into the atmosphere via improved waste management methods (Bengtsson et al., 2019). Furthermore, when combined with approaches such as networked climate governance which encourages collaboration among stakeholders in order to develop solutions tailored towards local contexts it becomes possible to better foster systemic change leading towards more sustainable outcomes over time (Winkels 2021).

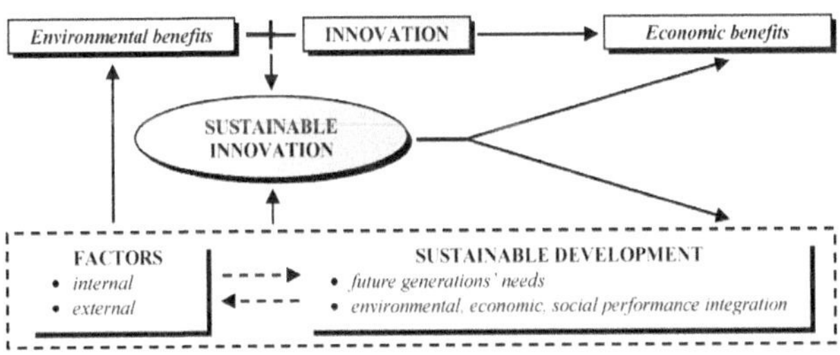

Figure 4.8. Example of a conceptualization of sustainable innovation.

Source: Pichlak and Szromek (2021).

Eco-innovation (Figure 4.8) provides a key complementarity here by enabling clean technologies that not only contribute directly toward improving environmental performance but also generate greater economic value adding potentials along with enhanced societal benefits overall (Von Weizsäcker et al., 2009). An example could come from geothermal heating/cooling systems which allow buildings around the world – including homes -to switch away from fossil fuels usage thus significantly lowering carbon footprints associated with burning these traditional fuels (IEA 2016).

Combining eco-innovative measures alongside initiatives within circular economies under collective climate action supported by networked climate governance makes it possible then to harness synergies between these two powerful frameworks providing us all with increased prospects for successful implementation of sound sustainability policy objectives moving forward (Gareau & Foerster 2020).

The third aspect focuses on the development and diffusion of eco-innovations and clean technologies that can reduce environmental impacts and support sustainable development (Bodin, 2017; Brandenberger et al., 2022; Rashchupkina, 2015; Van Popering-Verkerk et al., 2022; Wernli et al., 2023). Through networked climate governance, knowledge-sharing and technology transfer can be facilitated between governments, businesses, and other stakeholders to address challenges posed by climate change and promote the transition towards a circular economy. In recent years, eco-innovation and clean technologies have become increasingly important tools in the pursuit of a circular economy and sustainable development. As evidenced by the European Commission's 2018 report on the Eco-innovation Action Plan, such efforts are integral to establishing viable pathways towards climate action (EC, 2018). While there is still much work to be done in tackling climate change – including an urgent need for collective climate action – innovative solutions can assist governments around the world with meeting their goals. For example, one study found that investing in low carbon technology combined with policy support could reduce global emissions by up to 21 gigatons annually within 10 years (Held et al., 2019). Such advances demonstrate how renewable energy sources such as solar power and wind energy can help limit environmental impacts while providing reliable economic benefits for both businesses and consumers alike. Additionally, networked governance approaches which involve multiple stakeholders from different sectors has been identified as having potential benefits when it comes to creating successful strategies related to green innovation or achieving sustainability objectives (Braguinsky & Loorbach 2016; Ruggiero et al., 2017). The integration of eco-innovations into broader plans concerning sustainable development also holds promise when considering issues like disaster risk reduction or urban resilience initiatives. Climate change presents numerous challenges requiring adaptation measures designed specifically for local contexts – whether they focus on coastal areas threatened by rising sea levels or regions facing water scarcity due to changing precipitation patterns (Füssel & Klein 2014). In this sense, developing clean technologies tailored towards resilient cities may prove invaluable not only in terms of mitigating long term risks but also helping vulnerable populations better prepare themselves against more immediate threats. Overall then, investments in eco-innovative projects along with other forms of collaborative engagement remain essential if we wish to achieve success across all aspects of our current environment crisis–including those connected directly with climate resilience and adaptation needs.

The fourth aspect emphasizes the need for enhancing climate resilience and adaptation measures to minimize the effects of climate change on ecosystems, communities, and economies. Networked climate governance enables better coordination between stakeholders for developing and implementing effective policies, strategies, and infrastructure investments aimed at reducing vulnerability to climate risks and building resilience in a circular economy context. Climate resilience and adaptation are becoming increasingly important for a successful circular economy and sustainable development. To ensure effective collective climate action, networked climate governance must be utilized to build multi-stakeholder collaboration. According to a 2018 study by Smith et al., which analyzed the role of deliberative networks in managing environmental risks associated with industrial activities, "network governance is an emerging form of collaborative environmental management that involves multiple stakeholders" (p. 158). The author suggest that this type of governance can play a significant role in developing strategies aimed at increasing climate change resilience among social systems such as cities or countries. Moreover, they argue that "the success of network governance depends on engaging actors from all levels within government to collaboratively develop policies and approaches" (Smith et al., 2018, p. 159). This notion highlights the importance of policy integration when it comes to achieving meaningful progress towards greater sustainability through collective climate action and resilient outcomes under networked climate governance structures. To further illustrate how these concepts work together building upon each other for overall success; data collected from the Global Commission on Adaptations 2019 report shows that approximately $1 trillion dollars annually will need to be invested in global infrastructure alone over the next decade if we wish to meet current Sustainable Development Goals related to reducing poverty and inequality while simultaneously adapting successfully against impending threats posed by extreme weather events due largely in part caused by human induced Climate Change (Global Commission On Adaptation, 2019). Therefore, clearly emphasizing both the immediate necessity as well as long term benefits associated with investing heavily into proactive adaptive measures now before it is too late. This necessitates not only serious investment but also considerable effort across numerous sectors including business, finance civil society etc., working together collaboratively alongside governments governed via intelligent "networks" capable of facilitating efficient decision making processes so as maximize collective efficacy. It is also important to minimize wastefulness whilst optimizing public goods produced along the way thus effectively promoting more sustainable economic growth models within our societies.

The fifth principal aspect is the engagement of multiple stakeholders in collective climate action and networked climate governance (Anderton & Setzer, 2018; Bixler et al., 2023; Cody, 2018; Currarini et al., 2016; de Macedo & Jacobi, 2019). This involves connecting actors from different levels (local, regional, and global) and backgrounds (public, private, and civil society) to co-create solutions for sustainable development and circular economy transition. Policy integration across various sectors ensures that climate, economic, and social objectives are achieved in a coherent and complementary manner. Multi-stakeholder collaboration and policy integration have become essential aspects of circular economy and sustainable development in the pursuit of collective climate action. An example is seen in a 2018 article by Grießmeier et al., who highlight that "the implementation of the European Union's Circular Economy Action Plan (CEAP) requires an effective involvement of multiple stakeholders as well as adequate regulations, standards, policies and instruments such as economic incentives or public participation processes" (Grießmeier et al., 2018, p. 4).

Further evidence comes from a 2017 survey conducted by Kallis & Kovats which found that 93% of respondents agreed with the statement: 'multi-level governance approaches should be used to develop more integrated solutions for promoting sustainability at different levels' (Kallis & Kovats, 2017, p. 58). This sentiment has been echoed globally with nearly 200 countries signing on to The Paris Agreement – committing them all to pursue efforts towards limiting greenhouse gas emissions through networked climate governance initiatives. These multi-stakeholder collaborations are often combined with carbon markets where governments set caps on emissions allowing businesses to buy credits if they cannot meet their respective emission limits; creating financial incentive for companies transition away from high pollution activities while also generating revenue for government investments into green infrastructure projects like renewable energy sources or reforestation initiatives – furthering both global environmental protection strategies and sustainable economic growth objectives.

4.5. RESILIENCE AND ADAPTATION

Resilience and adaptation are important components of networked climate governance under the broad theme of collective climate action (Alexander et al., 2015; Davies et al., 2018; Den Uyl & Russel, 2018; Diop & Remvikos, 2016; Rashchupkina, 2015). Resilience is defined as the ability to withstand, absorb, adapt to, or rapidly recover from shocks or disturbances while maintaining essential functions and processes. Adaptation refers to adjustments in natural or human systems in response to actual or expected climatic stimuli and their effects. In order for resilience and adaptation strategies to be effective within a networked climate governance framework, there must be an emphasis on collaboration between different stakeholders such as governments, businesses, civil society organizations (CSOs), international institutions etc., who can work together towards common goals related to reducing emissions and adapting to changing climates. This requires strong leadership at all levels that encourages cooperation among actors with diverse interests. Additionally, it also necessitates adequate funding mechanisms so that resources can be allocated efficiently towards achieving these objectives. Finally, robust monitoring systems should be established in order ensure accountability amongst various stakeholders involved in this process.

Five aspects of resilience and adaptation in networked climate governance (McLaughlin, 2022; Medina-Garcia et al., 2022; Morrison et al., 2017). The first principal aspect of resilience and adaptation in networked climate governance involves a comprehensive assessment of the existing climate-related risks, vulnerabilities, and potential impacts (Figure 4.9). This stage also includes fostering awareness among stakeholders, policy-makers, and the general public about the importance of addressing these climate issues and the need for collective climate action. The concept of "resilience" has become increasingly important in the field of climate change governance, as it provides an opportunity to better anticipate and prepare for potential disruptions caused by changing environmental conditions (Zhang et al., 2019). Resilience refers to a system's ability to withstand and respond effectively to extreme events or pressures. However, there is still much work needed when it comes to assessing resilience levels and increasing awareness around adaptation strategies under networked climate governance. According to Rantala-Ylinen & Greco (2018), current assessment methods are often too simplistic and fail to capture the complexity of social-ecological systems. Additionally, many decision makers lack access or knowledge about how best use available resources for adapting their communities (Gonzalez et al., 2018). These issues with assessment and awareness can have significant implications on planning and prioritization efforts related to building resilient solutions that consider both human needs and ecosystem services (Levin et al., 2020). For example, without adequate understanding of which factors influence vulnerability at different scales within complex social-ecological systems – such as changes in land cover or regional economic dynamics – planners may prioritize interventions that provide short term gains but do not build long term adaptive capacity over time (Hinkel et al., 2017). This highlights the need for robust assessments tailored towards localized contexts so that effective policies can be developed; while also raising public awareness around these challenges through education initiatives designed specifically for each region's unique needs. Ultimately, successful implementation of comprehensive plans hinges upon having accurate information regarding the risks associated with various vulnerabilities– coupled with widespread understanding amongst stakeholders– so they are able to understand why certain actions must take priority over others based on expected outcomes resulting from specific decisions made today.

Figure 4.9. Example of various ways of improving climate risk assessments.

Source: Nature.

Once the assessment is complete and stakeholders are aware of the climate-related challenges, the next principal aspect is planning and prioritizing strategies and actions for resilience and adaptation. This might involve setting goals and targets, identifying the most vulnerable and affected sectors or regions, and determining the necessary resources to support these actions.

Climate change is a major challenge that necessitates the implementation of effective governance strategies to ensure meaningful adaptation and resilience. Networked climate governance has been identified as an appropriate approach for managing these complex problems (Füssel & Klein, 2018). Planning and prioritization are two essential components of this type of management; both must be taken into account in order to maximize effectiveness. Evidence from contemporary literature suggests that planning should focus on local-level risks, needs and resources associated with climate change impacts while also considering global targets such as those set out by the Paris Agreement (Gonzalez et al., 2020). This could involve developing risk maps which identify vulnerabilities at community level based on existing conditions or forecasting projections (Kemfert et al., 2019). Additionally, long term plans need to consider how socio-ecological systems may respond over time, taking into account potential changes related to human populations or technological innovations (van Oostenbrugge et al., 2018). Such comprehensive approaches can help inform policy design around how best to tackle issues related to climate change. Prioritization is another important element within networked climate governance. It involves determining what actions should take priority when faced with limited financial resources or capacity constraints due its complexity (Smit & Wandel 2006). For example, recent research found that urban areas around Europe have seen increased flooding frequency in recent years yet lack adequate protection against it – revealing a clear gap between perceived priorities versus actual investment decisions made by decision makers (Costa e Silva et al., 2019). To bridge this divide requires techniques like scenario analysis which weigh up various options before making any commitments concerning resource allocation (Pahl-Wostl 2008).

In doing so policies can create greater leverage towards achieving desired outcomes under constrained circumstances (Chaudhury et al., 2017; Chu, 2018; Cody, 2018). Overall then resilient networks require careful balancing of planning and prioritization processes if they are going to effectively manage the effects posed by anthropogenic induced climate change. While there will inevitably be challenges along the way, having evidence informed policy design coupled with rigorous evaluation criteria is key for ensuring equitable access amongst stakeholders whilst simultaneously furthering wider collective objectives – ultimately providing better opportunities for sustainable development in future generations (Huitema 2017). Through understanding the importance of proper planning and prioritization within networked climate governance frameworks we gain valuable insights regarding designing effective policies capable of facilitating adaptive responses commensurate with current realities relating to environmental degradation caused through anthropogenic activities worldwide. The success rate here depends heavily upon leveraging available data sources alongside ongoing assessments monitoring progress across multiple levels including individual communities all through national/subnational entities engaged in regional networking initiatives focused on sustainability goals. Therefore, being able evaluate resource availability combined findings gleaned from research projects studying social dynamics adds more clarity surrounding possible interventions required now meet emissions reduction standards consistently outlined international treaties like Paris Agreement thus equipping vulnerable populations living high risk zones means cope new threats confronting them increasingly unpredictable world impacted rapidly changing climatic patterns.

The third principal aspect focuses on designing policies and implementing measures that promote climate resilience and adaptation (Calliari et al., 2019; Deka & Goswami, 2022; Henry & Vollan, 2014; Krause et al., 2021; Saint Ville et al., 2016). These may include policies related to land-use planning, water and energy management, strengthening infrastructure, and enhancing social and economic capacities to address climate impacts. Effective policy design and implementation are key in ensuring that collective climate action and networked climate governance produce tangible results. Policy design and implementation in resilience and adaptation to climate change become increasingly important as the effects of global warming grow more severe. Networked climate governance, or a system of decentralized decision-making with multiple stakeholders involved in formulating policy (Gallagher & Knieling, 2018), can be effective in addressing these challenges; however, its success relies on robust policies that are designed thoughtfully and implemented carefully. Scholarly research has shown that integrated strategies for both mitigation and adaptation must be adopted if nations are to effectively manage their responses to climate change (Girardin et al., 2019). Such an approach requires collaboration across sectors such as energy, agriculture, finance, health care systems etc.; all leading towards developing comprehensive solutions at scale (Röckstrom et al., 2016). To achieve this level of integration—and thus build resilient communities against the impacts of global warming—it is essential for governments to invest resources into designing appropriate frameworks beforehand so that they may later implement them efficiently when needed (Lempert et al., 2017). The effectiveness of networked climate governance also depends upon flexible decision-making where adaptive learning plays a key role. Adaptive learning allows policymakers to adjust existing models based on knowledge gained from monitoring practices during program implementation (Moos & Lüdeke-Freund 2020); this enables authorities to identify gaps between predicted outcomes versus actual results which can then inform future decisions made concerning similar projects. Hence it is evident that flexibility alongside rapid feedback loops help ensure better overall performance through continuous improvement efforts over time by allowing dynamic responses according to changing circumstances throughout local contexts' environments (Dalhäuser & Koch 2019). Overall, it appears clear that policy design and implementation play substantial roles within the scope of resilience building under networked climate governance; however adequate attention should always be paid toward making sure measures are adapted flexibly so they may respond rapidly rather than rigidly while taking lessons learned from past experiences into consideration whenever possible. Only then will societies have any chance whatsoever at adequately responding swiftly enough before catastrophic damage becomes inevitable due modern day climatic changes already underway around our planet today.

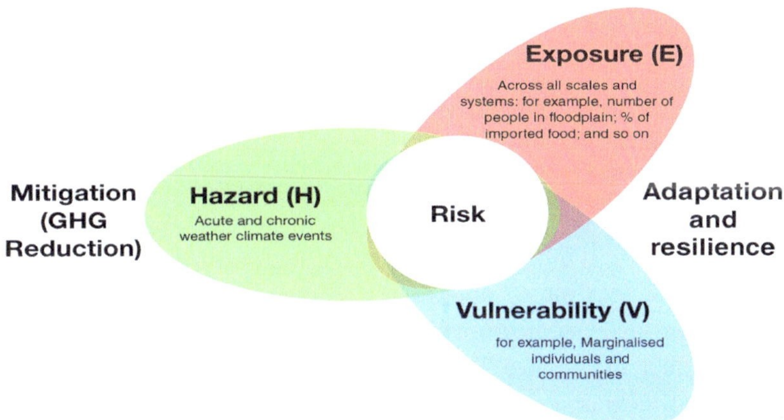

Figure 4.10. Understanding the dynamic nature of risk in climate change assessments.

Source: Royal Meteorological Society (RMetS) – Wiley.

The fourth principal aspect involves establishing an adaptive management framework that allows for continuous monitoring, learning, and adjusting of policies and strategies based on observed changes in climate and related impacts (Alexander et al., 2015; Anderton & Setzer, 2018; Archer, 2016). This requires a flexible decision-making process that can respond to new information and evolving needs, as well as a strong commitment to collaborate among diverse stakeholders in networked climate governance (Figure 4.10). The notion of establishing an adaptive management framework for continuous monitoring, learning, and modification of policies and strategies, grounded on the observed changes in climate and its consequent impacts, is pivotal. In this context, the Climate Change Response Framework (CCRF) developed in the northern forests of the United States serves as a salient case study. The CCRF, designed by the Northern Institute of Applied Climate Science, is perceived as a beacon of adaptation to climate change within the forestry sector. This ongoing project, initiated in 2009, underscores a holistic approach to adaptive management in response to aggravating climate change impacts on forest ecosystems. Its core elements encompass vulnerability assessment, education, and outreach, but most importantly, it integrates an adaptation toolkit that facilitates the development and implementation of forest management strategies in a changing climate. The toolkit employs a structured, step-by-step approach, from goal setting and assessment of potential impacts, through to implementation, monitoring and evaluation of adaptation actions. Trend-analysis forms the backbone of this flexible management; climate data is utilized not merely for predictive modeling but for tracking changes in forest composition and health over time to make necessary adjustments, ensuring the sustainability and resilience of the forests. Moreover, the CCRF is an example of networked climate governance with information shared across land managers, scientists, and professionals at local, regional and national scales through webinars, workshops, and publications. The potency of the CCRF allows a paradigm shift from a reactive stance to a proactive one, incurring vast policy implications, and rendering it possible to continually assess the trajectory and pace of climate change, and accordingly adapt forest management strategies in a timely and effective manner. By espousing such an adaptive management framework, policies and routines become dynamic instruments rather than static prescriptions, and can adjust to the fast-paced, unpredictable nature of climate change. Only then will societies have any chance whatsoever at adequately responding swiftly enough before catastrophic damage becomes inevitable due modern day climatic changes.

Lastly, resilience and adaptation in networked climate governance are heavily reliant on effective coordination, communication, and knowledge sharing among various actors within the governance network (Orach & Schluter, 2016; Ostrom, 2014; Panditharatne, 2016; Partelow et al., 2020; Pavanelli et al., 2023). Strengthening these connections is essential for overcoming barriers and fostering collective climate action at different levels, from local to global. This aspect also involves promoting transparency and accountability by sharing progress, successes, and challenges, which can further improve the overall effectiveness of networked climate governance. At the crux of resilience and adaptation in networked climate governance, effective coordination, communication, and knowledge sharing among various actors within the governance web are of high importance. The Great Barrier Reef Climate Change Adaptation Strategy and Action Plan (GBRCCASAP), established between 2012 and 2017, under the auspices of the Great Barrier Reef Marine Park Authority, epitomizes the pivotal role of these elements in achieving climate resilience. This comprehensive case study spanning across a six-year period offers substantial evidence of the interplay between a range of actors including local communities, Australian state and federal government bodies, eminent scientists, and traditional Aboriginal groups endeavoring collaboratively to manage and preserve this World Heritage site confronted by climate change threats. The effectiveness of the GBRCCASAP was fundamentally contingent on diligent coordination among the said entities, essential in synchronising climate adaptation efforts and dispelling risks of misalignment, duplications, or overlooked areas. It underscored profound dependency on robust communication channels, especially in simplifying complex scientific information about climate impacts into a digestible context, facilitating stakeholders' comprehension of associated risks and apprehension of adaptation measures, leading to proactive participation in the initiative. Further, the GBRCCASAP heightened its progress through fostering a knowledge sharing environment, providing a platform for actors to share research, experiences, and insights acquired over time that aided in crafting relevant, timely, and effective adaptation strategies. The success of the GBRCCASAP in augmenting resilience of the reef and its surrounding communities, therefore, is a resounding endorsement of the indispensability of coordination, communication, and knowledge sharing in any networked climate governance endeavour. In this light, these components are not just beneficial, but foundational for creating robust and versatile systems for climate adaptation.

CHAPTER 4 SUMMARY

Throughout this chapter, we have delved into the fascinating and intricate world of networked climate governance as a pivotal instrument in the quest for collective climate action. Exploring the fundamental principles, characteristics, and real-world applications of this paradigm, our analysis has not only underscored the potential of decentralized, collaborative approaches in addressing global climate challenges but also recognized the pitfalls and caveats that accompany such networks. As we conclude this chapter, it is essential to reflect upon the insights gleaned and consider how they might inform the direction of future climate action. To recapitulate, the limitations of traditional, centralized governance models in addressing complex, global-scale issues such as climate change have necessitated the emergence and growing relevance of networked climate governance. As evidenced by various examples discussed in this chapter, these networks foster synergies by bringing together diverse actors ranging from national, subnational, and local governments, supranational organizations, private enterprises, non-governmental organizations, and civil society actors. Moreover, the principles of inclusion, adaptability, and polycentricity inherent in networked governance have demonstrated their potential to enhance the effectiveness, responsiveness, and resonance of climate policy instruments and strategies.

Nonetheless, it is crucial that we remain cognizant of the challenges that networked climate governance may face. Our analysis has delineated potential pitfalls associated with these networks, such as power imbalances, expertise disparities, and inefficiencies arising from overlapping memberships. To maximize the impact of networked governance efforts, policymakers and stakeholders must continue to strive towards addressing these obstacles and refining their collaborative approaches. Moving forward, further research and exploration of networked climate governance will be instrumental in understanding its long-term implications on global climate action. Moreover, it will be valuable to study the diffusion and cross-pollination of knowledge, innovation, and best practices facilitated by these networks, thereby fostering learning-by-doing and the continual improvement of climate policies across different jurisdictions. Lastly, fostering transparency and accountability within networked governance arrangements will remain paramount in ensuring that these collaborative efforts effectively safeguard the environment and drive the progression towards a sustainable future.

In conclusion, networked climate governance presents a promising avenue for promoting collective climate action in the face of unprecedented global interdependence and environmental challenges. The interconnectedness, adaptive capacity, and decentralized nature of this governance model make it well-suited to address the complexities and uncertainties of climate change. Yet, it is essential for stakeholders to remain vigilant against the challenges posed by power structures, expertise disparities, and potential inefficiencies, to ensure the efficacy of their collaborative efforts. By doing so, networked climate governance can serve as an invaluable instrument in our collective pursuit of a greener, more sustainable world. Having explored the dynamics and implications of networked climate governance, it is crucial to examine other policy instruments and frameworks employed in addressing climate change. One such mechanism is the market-based approach, specifically carbon markets and emissions trading, which offers an economic incentive to reduce greenhouse gas emissions. In the following chapter, we shall delve into the intricacies of these market-based mechanisms, analyzing their potential, shortcomings, and role within the broader context of global climate governance. By exploring the interconnections between networked governance and market-based approaches, we will gain a more comprehensive understanding of the multifaceted, collective strategies required to tackle the global climate crisis.

CHAPTER 5

Carbon Markets and Emissions Trading

CONTENTS

As we delve further into the interconnected realms of collective climate action and networked climate governance, it becomes essential to shine a spotlight on one of the most prominent policy instruments for addressing climate change – carbon markets and emissions trading. In this fifth chapter of the book, we will examine the fundamental structures underpinning these market-based mechanisms, explore prominent examples in operation across the globe, and evaluate their efficacy in augmenting networked climate governance approaches.

The primary premise of carbon markets is to provide financial incentives to companies and organizations that adopt environmentally sustainable practices to reduce their emissions (Dey et al., 2017; Diop & Remvikos, 2016; Ensor & Harvey, 2015; Fidelis et al., 2019). Emissions trading, also known as "cap-and-trade," plays a critical role in establishing these incentives by placing an upper threshold on the quantity of greenhouse gas emissions permissible within a specific region, industry sector, or community. Those who emit above their allocated limits are required to purchase emission allowances (carbon credits) from entities that emit below their designated quotas. In essence, the cap-and-trade system empowers industries and organizations to either intensify their decarbonization efforts or bear the financial implications of exceeding their emission allowances.

A central aspect that we must unravel in our discussion of carbon markets is the linkage between their operational design and the ultimate goal of mitigating anthropogenic greenhouse gas emissions. The globally ratified Paris Agreement enshrines the objective of limiting global temperature rise to 1.5°C above pre-industrial levels. Accordingly, we will scrutinize how cap-and-trade regimes enable countries and sectors to achieve their Nationally Determined Contributions (NDCs) in line with this shared climate target. Given this context, we will delve into the potential harmonization and collaboration among carbon markets, which have the potential to facilitate international climate cooperation and widen the scope of national decarbonization actions.

As we embark on an analytical journey through the world of carbon markets, it is imperative to examine a variety of case studies that underscore their implementation and impact. We will recount the inception and evolution of the European Union's Emissions Trading System (EU ETS), the largest and most comprehensive carbon market in operation today. Additionally, we will investigate various emerging and established carbon markets in jurisdictions such as North America, Asia, and Australia to identify common challenges, best practices, and innovative mechanisms that have the potential to inform and enhance ongoing global efforts in emissions trading.

An exhaustive analysis of carbon markets and emissions trading would be incomplete without addressing the pitfalls and shortcomings associated with these practices. In this chapter, we will also candidly confront the concerns and criticisms levied against carbon markets, including the allegations of perpetuating environmental injustice, disproportionately burdening developing countries, and presenting opportunities for fraudulent activities. Furthermore, we will shed light on the technical complexities and potential inefficiencies encountered in the design and implementation of emissions trading schemes, including carbon credit volatility and allocation methodologies.

In conclusion, this chapter aims to provide a holistic and balanced perspective on carbon markets and emissions trading in the broader context of collective climate action and networked climate governance. By demystifying these policy instruments and presenting critical insights, we hope to contribute to an enriched understanding of the interplay between climate policy, international cooperation, and sustainable development. Ultimately, it is our aspiration to enrich the ongoing discourse and foster a more nuanced appreciation for the challenges and opportunities presented by carbon markets and emissions trading in the quest for tangible climate change mitigation.

5.1. THE KYOTO PROTOCOL

The Kyoto Protocol is an international agreement that was adopted in 1997 to reduce global greenhouse gas emissions. It sets binding targets for industrialized countries to reduce their emissions of six major greenhouse gasses, including carbon dioxide (CO_2). The protocol also established a system of "emissions trading" which allows countries to buy and sell emission allowances with each other. This has created a market-based approach to reducing emissions by providing incentives for companies and individuals to invest in clean energy technologies. The Kyoto Protocol is considered one of the most important components of the global effort towards tackling climate change as it provides a framework for nations around the world to work together on reducing emissions (Friedman et al., 2020; Niles & Lubell, 2012; Tosun & Schoenefeld, 2017; Tuda & Machumu, 2019). Under this agreement, participating countries are required to meet specific commitments regarding their total annual CO_2 emissions over specified periods of time.

These commitments can be met through various means such as investing in renewable energy sources or implementing policies that promote efficient use of resources like electricity and fuel consumption. In addition, developed countries have agreed to provide financial assistance and technology transfer support so developing nations can better manage their own GHG reduction efforts. Overall, the Kyoto Protocol serves as an essential component within carbon markets and emissions trading systems due its ability to create economic incentives for businesses while simultaneously helping protect our environment from further damage caused by human activities related to climate change.

The Kyoto Protocol was groundbreaking in that it established a system of emissions trading, which allowed countries to buy and sell emission allowances as part of their effort to reduce greenhouse gas (GHG) emissions. According to the World Bank's Carbon Pricing Dashboard, over 40 jurisdictions around the world have implemented some form of carbon pricing policy or are planning on doing so by 2020 (World Bank Group, 2018). The Emission Trading System (ETS), created under the protocol's framework, is an example of one such measure designed with the intention of incentivizing GHG reduction at both national and international levels. This kind of market-based approach makes it possible for countries who cannot meet their own mitigation targets due to economic constraints to purchase emission credits from other nations who can exceed theirs; thus allowing them all collectively reach overall goals set out by the Protocol while also providing additional financial support for those facing greater difficulties (Bodansky & Moomaw 2008). Moreover, this mechanism has proven highly efficient in facilitating cost reductions associated with compliance since participants benefit from price signals generated through competitive markets rather than relying solely on centralized government regulation alone (Tavoni et al., 2014). Additionally, studies suggest that ETS has had positive effects in terms of stimulating investments into clean technologies research and development as well as spurring innovation across participating regions where prices tend to fluctuate more often than average due their frequent participation in larger cap-and-trade systems like European Union Emissions Trading Scheme (EU ETS) which is illustrated in Figure 5.1 (Chen et al., 2017).

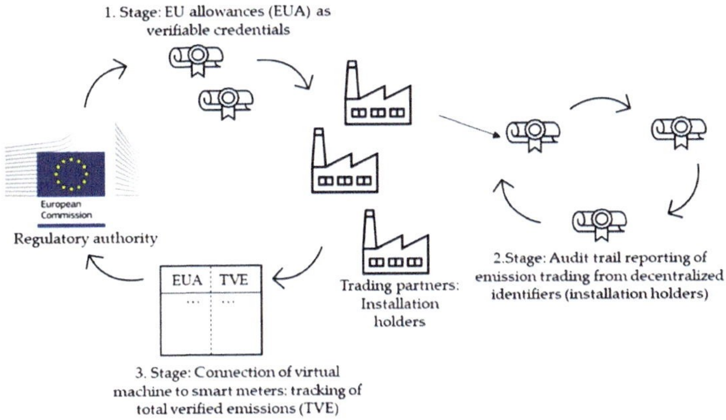

Figure 5.1. The three working stages in our identity layer for the European Union Emissions Trading System (EU ETS).

Source: Mandaroux et al. (2021).

The Kyoto Protocol, adopted in 1997 and entered into force in 2005, created the Clean Development Mechanism (CDM), which allows industrialized countries to invest in projects that reduce emissions in developing countries while receiving credits for those reductions. Since its inception, more than 8500 CDM-registered projects have been approved by the United Nations Framework Convention on Climate Change (UNFCCC) across 129 different host parties (Grecco & de Freitas Gomes, 2018). These projects span a range of activities from renewable energy production to methane capture from agricultural waste management operations (Wania et al., 2019). The effectiveness of the CDM has been widely discussed and evaluated since it was established under the Kyoto Protocol. A study conducted by Elshawy et al. (2018) found that higher levels of investment led to greater emission reduction gains compared to areas with lower investments; however, there remains considerable debate as to whether or not these benefits are sufficient enough for overall environmental gain given their cost and complexity relative other policy options available today. Similarly, another assessment revealed that although efforts were made towards sustainable development objectives such as poverty alleviation and improved access to electricity through project funding mechanisms supported by the CDM mechanism had positive effects at an individual level, they did little when looked at cumulatively on broader scales due community displacement caused by large scale infrastructure projects related to registered initiatives (Gomez-Echeverri et al., 2020). Additionally, researchers have raised concerns about problems associated with double counting whereby carbon offsets generated via CDMs could be claimed multiple times resulting in further skewing our understanding of true mitigation potentials if accurate tracking measures are not put into place (Scott & Wurster, 2019). Overall, we can see that even though some progress has been made regarding reducing emissions through the implementation of various clean development programs funded under the guise of credits offered by The Kyoto Protocol's CDM initiative; much work is still needed including proper monitoring systems and better accounting methods being employed if any meaningful long term benefit is going too achieved.

The Kyoto Protocol, which was adopted in 1997 and entered into force on February 16th 2005 (United Nations Framework Convention on Climate Change [UNFCCC], 2018), set up JI projects by developed countries as a way to achieve emissions reduction commitments. These JI projects are designed for two or more parties from Annex I countries of the UNFCCC to cooperate and share credits resulting from their reductions efforts, with one party receiving credit for its effort towards meeting emissions targets while other party involved benefits economically from participation in the project (Kirilenko & Oberndorfer, 2016). The overall purpose of these JI projects is to reduce costs associated with achieving emission limits through international cooperation among governments and businesses (Golan et al., 2019). Since it became an option under the Kyoto Protocol in 2001, there has been significant growth in investment related to such joint implementation (JI) activities— rising 3-fold between 2000 and 2012 globally according to United Nations data released in 2013 (as cited Kornova & Ollarves, 2015). This increase can be attributed largely due increasing number of major emitting nations joining the protocol over that period including: Japan; Russia; Canada; New Zealand; Belarus; Ukraine; Kazakhstan, Hungary, Iceland, Latvia, Lithuania, Monaco, Slovakia, Slovenia, Romania, Bulgaria, Croatia, Estonia, Macedonia, Liechtenstein, Serbia, Bosnia Herzegovina, Armenia, Azerbaijan, Georgia, Moldova Republic of South Ossetia, Montenegro, Albania, Holy See, San Marino, Andorra, Palestine, Faroe Islands, Vatican City State, Northern Cyprus, Cook Islands, Niue, Tuvalu, Marshall Islands, Palau, Solomon Islands, Micronesia, Nauru, Tonga, Kiribati, Vanuatu, Federated States Of Samoa, Tokelau, Wallis, Futuna, American Samoa, Anguilla, British Virgin Island, Montserrat, Turks Caicos, Sint Maarten, Pitcairn, Saint Helena Ascension, Tristan Da Cunha, Bermuda, Guernsey, Isle Man Jersey, U.S Minor Outlying islands, Bonaire, Saba, Saint Eustatius, Greenland, French Polynesia, Netherlands, Antilles,

East Timor, Macao, China, Hong Kong, and the Special Administrative Region of Taiwan. Yet, the USA has not always been an enthusiastic partner in these projects.

The United States disengagement from the Kyoto Protocol carries several implications that affect both the nation and the international community (Alexander et al., 2015; Brewer et al., 2017; Ensor & Harvey, 2015; O'Brien et al., 2021; Perz et al., 2022). The Kyoto Protocol, adopted in 1997, aimed to reduce greenhouse gas emissions worldwide, with particular focus on developed countries which emit a significant portion of global greenhouse gasses.

1. *Environmental Impact*: The US disengagement from the Kyoto Protocol raises concerns over the potential increase in greenhouse gas emissions. The lack of commitment to reduce emissions could lead to long-term environmental consequences such as climate change and increased air pollution, along with the associated adverse effects on human health. Furthermore, the US's withdrawal might impact the willingness of other countries to engage in climate change initiatives or uphold their own commitments, potentially having a multiplying effect on global emissions.
2. *Economic Impact*: By not participating in the Kyoto Protocol, the US may lose access to certain international carbon markets, affecting the competitiveness of its businesses, especially those involved in clean technologies. This could result in investment opportunities shifting towards other nations that are actively participating in climate change mitigation efforts. In addition, the costs associated with climate change impacts, such as increased severity and frequency of natural disasters, could challenge the US economy.
3. *Political Impact*: The US disengagement sends a message to the international community, potentially undermining global efforts to address climate change. This might hamper diplomatic relations and cooperation between the US and other countries in global environmental and sustainability efforts. Additionally, the withdrawal could create a power vacuum, enabling other nations to take up leadership roles in the fight against climate change.
4. *Influence on Private Sector*: The absence of binding emission reduction commitments for the US under the Kyoto Protocol might

influence businesses in the country to avoid making necessary investments in low-carbon and sustainable technologies, as there is less regulatory pressure. This could lead to a slower shift towards cleaner energy sources compared to other countries committed to the Kyoto Protocol.

5. *Legal Implications*: By disengaging from the Kyoto Protocol, the US is not legally bound to meet specific emission reduction targets. This may result in a lack of clear environmental regulations and enforcement mechanisms, which could give way to increased greenhouse gas emissions in the country. Overall, the implications of US disengagement from the Kyoto Protocol are multifaceted and bear significant consequences. While the impact of this decision on global efforts to combat climate change and align the world towards a more sustainable future is uncertain, it highlights the importance of collective efforts and cooperation among nations to address the pressing issue of climate change.

International Emissions Trading (IET) is a mechanism within the Kyoto Protocol that allows countries to meet their emissions reduction targets by trading or purchasing allowances for GHG emissions (Granderson, 2017; Gregg et al., 2023; Gronow et al., 2020). Under IET, developed nations with higher emission levels can purchase credits from developing nations who have achieved reductions in greenhouse gasses below their assigned target limits (Carrus et al., 2019).

This type of market-based approach helps these countries reach targets more quickly and cheaply than through other methods such as regulation or setting up carbon taxes on industry. Furthermore, it encourages those countries selling the credits to increase investment in renewable energy sources & low-carbon technologies since they are able to receive economic compensation for doing so (Bondarenko, 2018).

The Adaptation Fund Projects provide an important link between International Emission Trading and climate change adaptation. The funds generated through international markets help finance projects designed specifically to address impacts associated with climate change which may otherwise be difficult to develop due lack of financing options available at local level (Linke & Krenzlin, 2018). For example, one recent project funded by the Adaptation Funds was used to construct flood barriers near Bangladesh's capital city Dhaka thereby protecting over 8 million people living in nearby areas from flooding disasters caused by melting glaciers and rising sea levels (Tripathi et al., 2018). Such initiatives would not be possible without access to financial resources generated through successful implementation of IET mechanisms under Kyoto Protocol (Figure 5.2).

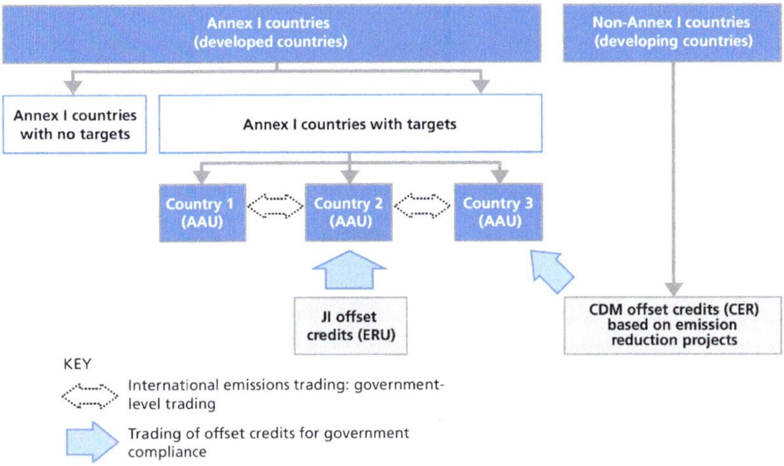

Figure 5.2. The UNFCCC carbon trading mechanisms.

Source: Lazarowicz (2019).

The Adaptation Fund is a financial mechanism established under the Kyoto Protocol to help developing countries adapt to climate change (Bernstein & Hoffmann, 2018; Koefoed, 2019; Lubell & Robbins, 2022; Sara et al., 2016; Therville et al., 2019). It provides direct funding for projects and programs that build resilience, reduce vulnerability, and enhance adaptive capacity in vulnerable communities around the world (UNFCCC 2020). These funds are generated from two sources: contributions made directly by governments or entities associated with them; and proceeds from emissions trading of Annex 1 Parties subject to emission limitations commitments (Kopetz et al., 2018). Projects funded through this fund must meet certain criteria such as being "cost-effective," having clear objectives, and showing measurable results on a local level (Harrison & Loveluck 2019). The goal of these projects is to improve climate adaptation efforts within developing countries while helping them move towards sustainable development pathways (Meyer et al., 2017). They can range from small scale initiatives focused on building resilient infrastructure, improving agricultural practices, strengthening early warning systems etc. One example would be an initiative implemented by India which aimed at promoting better energy efficiency leading to reduced greenhouse gas emissions resulting in improved air quality across Delhi's schools (Arora et al., 2019). This project was able to greatly increase public awareness about the need for energy conservation measures among school children thereby reducing overall carbon footprint in urban areas.

These funds have been instrumental in providing much needed resources for many low-income nations who otherwise lack access to capital markets due to their limited economic means or simply because they do not have enough experience dealing with international finance mechanisms like carbon markets or emissions trading schemes (Wyborn, 2015; Yates, 2012; Yoder et al., 2022; Zeren et al., 2022; Zhan, 2022). While successful implementation of Adaptation Funds has helped mitigate some risks posed by climate change it still remains one component of transitioning economies away from reliance on fossil fuels. In this regard the European Union Emission Trading System serves as a more prominent tool where large emitters are required to comply with stringent limits set forth under EU law thus ensuring greater compliance rates than what could be achieved without legal framework governing CO_2 reduction targets set out therein (EUETS Fact Sheet 2020). Hence we see how both tools work together synergistically so as to ensure long term environmental sustainability goals become reality over time rather than short lived successes here today gone tomorrow approach taken often when only relying upon individual solutions.

5.2. THE EU EMISSIONS TRADING SYSTEM

The EU Emissions Trading System (ETS) is a carbon market and emissions trading system that was established in 2005. It is the world's largest cap-and-trade program, covering more than 11,000 power plants and factories across Europe (Willems & Giezen, 2022; Wyborn, 2015; Yates, 2012; Yoder et al., 2022; Zeren et al., 2022; Zhan, 2022). The ETS works by setting an overall limit on greenhouse gas emissions from these facilities, which are then allocated to each facility as allowances. Facilities can buy or sell their allowances depending on whether they need additional ones or have surplus ones available. This creates a financial incentive for companies to reduce their emissions since it costs them money if they exceed their allowance limits. The goal of the ETS is to help European countries meet their commitments under the Paris Agreement and other international climate agreements by reducing greenhouse gas emissions cost effectively while creating economic opportunities for businesses involved in low-carbon activities such as renewable energy production.

The European Union Emissions Trading System (ETS) is a flagship policy instrument of the European Union to reduce greenhouse gas emissions and combat climate change (Goodwin et al., 2022; Gordon, 2016; Gore, 2010). It works by putting an upper limit, or cap, on the total amount of carbon dioxide that can be emitted by all participating installations in the EU ETS trading system. Companies must hold allowances for each ton of CO_2 they emit, which are allocated via national governments using either auctions or free allocation. If companies exceed their allowance limit then they need to purchase additional permits from other participants who have not used up theirs ("secondary market"). This creates incentives for businesses to invest in new technologies and processes that will help them stay below their emission limits while also generating revenue through the sale of unused allowances ("primary market"). The changes in this market are almost always linked to geopolitical events (Figure 5.3). Nevertheless, there are impressive statistics to support the efficacy of the program.

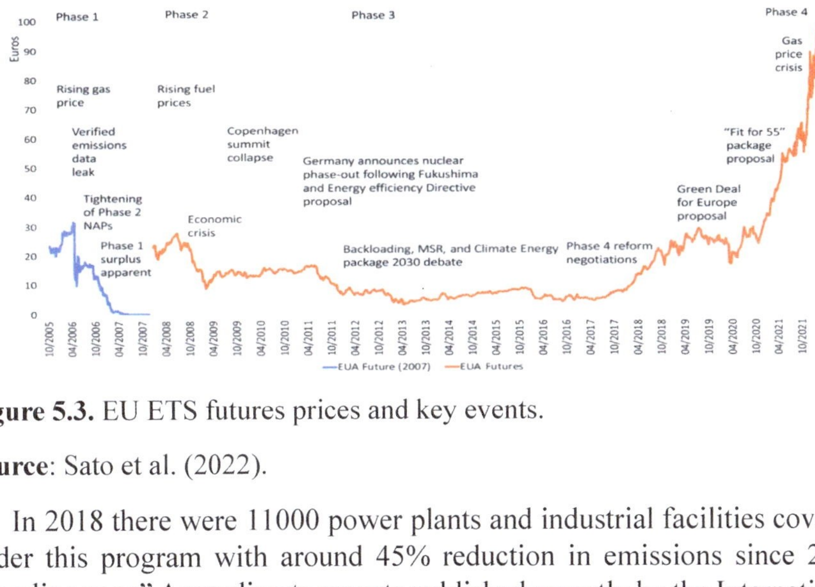

Figure 5.3. EU ETS futures prices and key events.

Source: Sato et al. (2022).

In 2018 there were 11000 power plants and industrial facilities covered under this program with around 45% reduction in emissions since 2005 "baseline year." According to reports published recently by the International Energy Agency (IEA), global carbon dioxide emissions rose 1.7% in 2018 despite growing investment towards renewable energy sources such as windmills and solar panels across Europe. These results further emphasize the importance of effective implementation of policies like EU ETS if we want progress toward achieving our internationally agreed-upon goals set out during Paris Agreement 2015. Furthermore, recent research conducted at Stockholm Environment Institute has shown how net economic gains ranging between €18 billion – €24 billion could be expected over 10 years period due to reduced pollution levels associated with successful implementation of these systems which should incentivize more countries worldwide to consider adopting similar approaches within their own jurisdictions too (Hilaire et al., 2019).

The European Union's Emissions Trading System (EU ETS) is the world's largest carbon market, implemented in 2005 to reduce greenhouse gas emissions from sectors such as power generation and manufacturing (Niles et al., 2021; O'Brien et al., 2021; Olivier & Berardo, 2022). Under this system, companies must hold allowances for each ton of CO_2 they emit. These allowances are tradable credits that give their holders permission to release a certain amount of carbon into the atmosphere; if an entity exceeds its allowance then it has to purchase additional permits from another company with fewer emissions than their own quota allows them. This encourages companies to invest in clean energy initiatives and technologies to help reduce their overall emission levels or else face having to buy more expensive permit credits on the open market. In order for these EU ETS regulations to be effective however there needs to be confidence amongst stakeholders in both the accuracy and reliability of reported data by member states (Gill et al., 2018). To ensure this happens, national authorities have been given significant responsibility over monitoring, verifying and enforcing compliance—which includes taking corrective action against participants who do not comply with reporting requirements or those found manipulating information about their activities (OECD/ITF & International Carbon Action Partnership [ICAP], 2019). The effectiveness of these measures can also depend on other factors such as availability of resources at a national level which may vary between countries depending upon budget constraints or lack thereof (Briggs & Newell-Price, 2017). In addition, penalization systems should be robust enough so that potential noncompliant entities understand what actions will result in sanctions being applied but without creating unfair economic burdens for legitimate businesses seeking access via the EU ETS marketplace. Consequently when considering how best Companies might address any responsibilities under carbon markets and emissions trading within the EU ETS framework special attention needs paid towards understanding all relevant aspects involved including accurate records keeping procedures together with appropriate sanctioning mechanisms designed specifically promote positive behavior whilst simultaneously deterring negative practices where possible – thus allowing firms greater certainty regarding expected outcomes associated with participating either directly through buying offset credits or indirectly via investing cleaner energy sources.

The EU Emissions Trading System (EU ETS) is the cornerstone of Europe's climate policy, and has been in place since 2005 as a way to reduce greenhouse gas emissions. Companies covered by the system are allocated or auctioned allowances for each ton of CO_2 they emit, which can be traded between companies to help them meet their obligations under the scheme (Chaudhry et al., 2018). This trading mechanism provides an opportunity for companies to purchase emission allowances from other entities that have surplus permits; this allows those with high-emission activities to 'offset' their levels through buying credits elsewhere within the market. According to Chaudhry et al. (2018), studies suggest that ETS enables cost efficient reductions in carbon dioxide equivalent emissions up until 2020 compared with non-market based measures such as command and control regulation due its flexibility in allowing participants multiple ways in achieving compliance targets at lowest possible costs. Therefore, it appears more beneficial than traditional regulatory approaches when setting limits on pollution levels across industries (Chen & Dubin, 2019). Furthermore, research conducted by Chen & Dubin (2019) indicates that there is also potential for introducing offsets into international markets outside of the European Union – namely via JI projects and CDMs developed under Kyoto Protocol – allowing eligible organizations located anywhere around world who invest in approved low-carbon energy technologies or renewable energy sources access to offsetting opportunities available through these schemes. As mentioned previously this could potentially benefit firms operating within domestic carbon markets if used strategically given most countries lack substantive national policies addressing emissions reduction goals over long term (Heldenbrand et al., 2017). Overall, it seems clear that not only does participation in EUETS provide options for reducing overall company level emissions but also offers businesses greater scope beyond domestic borders enabling them to find practical solutions in sectors where meeting regulations may otherwise prove difficult or costly to achieve independently (Figure 5.4 for an illustration of this phenomenon in the aviation industry).

The European Union Emissions Trading System (EU ETS) has been established to reduce greenhouse gas emissions in the region through carbon markets and emissions trading. Companies are able to offset their own emissions by investing in non-EU projects that also work towards reducing global warming (Bixler et al., 2023; Blok, 2018; Orach & Schluter, 2016; Teodoro et al., 2021; Watabe & Gilby, 2020). This mechanism is known as 'backloading,' which involves buying up allowances from other countries or organizations for use within the EU's cap on total allowable emission levels.

Recent research conducted by Zang et al (2018) found that backloading was successful in providing a more uniform price level across different regions of Europe, while still maintaining overall environmental objectives set out by the EU ETS system.

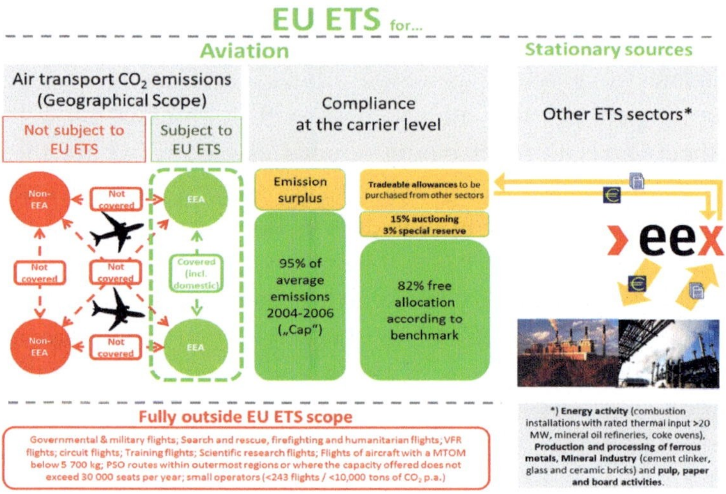

Figure 5.4. Design of the EU Emissions Trading Scheme (EU ETS) for aviation.

Source: Maertens et al. (2019).

The study concluded that this approach effectively enabled companies to establish an effective cost-benefit analysis when evaluating investment options outside of the EU. Additionally, it allowed these businesses access to cheaper alternatives than those available domestically whilst still achieving their climate goals at home. In addition, case studies have demonstrated several examples where large corporations such as Microsoft Corp., Unilever NV and Google Inc., have successfully implemented backloading into their strategies for reducing their respective CO_2 footprints (Gosling et al 2018). For example, Google invested approximately \$50 million USD into Indian wind farms during 2015; thus purchasing renewable energy certificates from India with corresponding tradable credits under the International Carbon Action Partnership framework (ICAP). In doing so they were able to meet certain domestic requirements while taking advantage of lower prices abroad compared with what would be offered within Europe itself. Finally according to data compiled by Eurostat between 2013 – 2017 there has been consistent growth year over year in terms of investments made into external carbon reduction initiatives; rising from €1 billion euros per annum initially up to nearly €3 billion euros per annum most recently indicating strong corporate support for alternative measures beyond traditional compliance mechanisms provided only locally via government bodies like Environment Canada and EPA. Overall, given its efficacy since its implementation along with increasingly widespread acceptance among big business looking for better value elsewhere – Backloading appears poised continue being leveraged both now and well into future as part of comprehensive strategy when addressing pollution related issues particularly throughout Europe but potentially further afield too depending upon circumstance & demand alike moving forward just same as ever before.

Backloading is a mechanism which enables governments to temporarily withhold emissions allowances in the EU Emissions Trading System (EU ETS), under carbon markets and emissions trading (Koontz et al., 2015; Krause et al., 2021; Laycock & Mitchell, 2019; Lee et al., 2014). This mechanism has been used extensively since it was introduced in 2013, most notably during the first phase of its implementation from 2014–2019. Backloading involves withholding a portion of emission permits that would otherwise be auctioned off at regular intervals on an annual basis; instead these are "banked" for later use when demand increases or prices rise due to scarcity over time. The purpose of backloading is to maintain price stability while also providing financial support for low-carbon investments such as renewable energy sources. The Clean Development Mechanism (CDM) provides another avenue by which countries may achieve their climate change commitments as part of international agreements like the Kyoto Protocol and Paris Agreement, through projects aimed at reducing greenhouse gas emissions abroad. CDM allows developed countries with binding targets under those agreements to invest in sustainable development projects located within developing nations, often receiving credits toward meeting their own legal obligations as well as generating additional revenue streams for the host country itself – all while helping reduce global GHG levels overall. As both mechanisms—backloading and CDM—enable governments to meet their environmental goals financially without sacrificing economic growth potential, they have become increasingly popular options among many countries looking towards greener futures.

5.3. THE CLEAN DEVELOPMENT MECHANISM

The CDM is a part of the carbon markets and emissions trading system (Luthe & Wyss, 2016; Salmoral et al., 2020; Vantaggiato et al., 2023;

Watabe & Gilby, 2020). It was established under the Kyoto Protocol in 1997, and it allows industrialized countries to invest in projects that reduce greenhouse gas emissions in developing countries. The CDM provides an incentive for developed countries to invest in clean energy technologies by allowing them to earn credits which can be used towards meeting their own emission reduction targets. Under the CDM, companies or governments from developed nations are able to purchase Certified Emission Reductions (CERs), which represent one ton of CO_2 equivalent reduced through approved projects located within developing nations. These CERs can then be sold on international carbon markets as tradable commodities, providing financial incentives for investments into low-carbon development activities such as renewable energy generation or improved efficiency measures.

The CDM is an international mechanism created by the Kyoto Protocol to help reduce greenhouse gas emissions (Bailey et al., 2012; Bodin et al., 2019; Grundig & Ward, 2015; Gunningham, 2012; Vasconcelos et al., 2015). It encourages developed countries to invest in emission-reducing projects, such as renewable energy and energy efficiency improvements, in developing countries in order to receive credits for their reductions achieved. This system allows industrialized nations that have agreed to comply with emissions targets set forth by the UNFCCC – including industrial giants like China, Japan, and Germany – to trade carbon dioxide reduction allowances between themselves or other UNFCCC signatories. As a result of this arrangement, large corporations are able to fund projects abroad while simultaneously reducing their own domestic emissions profiles without any significant immediate costs (Swanson & Hultman 2012). In addition, CDM has proven economically efficient compared with traditional methods of achieving GHG mitigation goals because it incentivizes companies "to adopt least-cost approaches for project implementation" due its flexible approach towards compliance options and access points available at different stages of development (Lee et al., 2019). For example, according to data from 2018 provided by the World Bank Group's Carbon Finance Unit (CFU), over 2200 registered CDM projects around the world had generated more than 1 billion Certified Emission Reduction units since 2005 — representing annual savings of approximately 790 million metric tons of CO_2e per year globally during that period (World Bank Group 2020). Therefore, not only does it provide environmental benefits but also economic ones through cost effectiveness when implemented correctly. Overall, The CDM provides an effective way for developed countries who are committed under the UNFCCC agreement to find ways to support climate change initiatives overseas while receiving credit back home; thus providing both environmental and economic incentives which make up a win-win situation overall.

The Kyoto Protocol and Paris Agreement are examples of global climate action agreements that set emissions reduction targets for participating nations (Anderton & Setzer, 2018; Kim et al., 2019; Olivier & Berardo, 2022; Vantaggiato et al., 2023; Yoder et al., 2022). These agreements provide mechanisms such as carbon credits which allow investing countries to offset their own greenhouse gas emissions by providing funding or technology transfer initiatives to other, developing countries (Rose et al., 2013). Carbon credits (Figure 5.5) can be used in the trading markets created under these international agreements; individual investors purchase them from another country with lower emission levels than themselves, allowing the purchasing nation to meet its domestic requirements outlined in the agreement (Munoz et al., 2011).

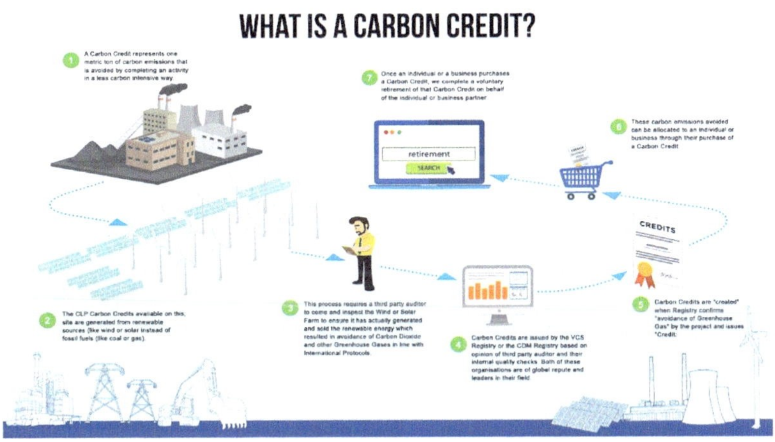

Figure 5.5. Conceptualization of a carbon credit.

Source: CLP Carbon Credits.

For example, when China signed on to both The Kyoto Protocol and Paris Agreement they were given access to a CDM where Chinese companies could buy offsets from projects located abroad at a preferred rate compared to those traded domestically within China's market-based systems (Ohara & Lipton, 2018). Through this mechanism China was able to reduce over 2 billion tons of CO_2 equivalent since 2006 through foreign investments alone – significantly more than if relying solely on domestic strategies without CDMs (Liu et al., 2019). This demonstrates how effective using carbon credits can be in helping an investing country meet their own emissions reduction commitments while supporting sustainable development overseas. Overall, carbon credits offer an efficient way for developed countries like China who have agreed upon collective climate action treaties such as The Kyoto Protocol or Paris Agreement to achieve their national target goals relating to reducing harmful emissions. By offering financial incentives and technological support internationally via projects based outside of their borders, these developing countries can use carbon credits to make up any shortfall in achieving their obligations under the treaty and help contribute towards mitigating global warming on a worldwide scale.

In order to effectively demonstrate environmental integrity and provide additional benefits beyond those of other mitigation activities, CDM projects must adhere to certain standards (Cody, 2018; Currarini et al., 2016; Fitch-Roy et al., 2020; Hawkins et al., 2018; Tuda & Machumu, 2019). Studies have found that the implementation of these high-quality CDM projects can result in positive outcomes such as improved air quality, increased access to clean energy sources, reduction of greenhouse gas emissions, conservation efforts for biodiversity and ecosystems, job creation opportunities within host countries or regions where the project is located (Chen et al., 2018; Kojima & Sato 2008). Statistics from an analysis conducted by UNEP Risoe Centre show that between 2003–2009 there was a total estimated GHG emission reduction equivalent of 1.2 gigatons CO_2 through successful Certified Emission Reduction Projects under the CDM (Snyder & Miranda 2011). Furthermore, according to research done on renewable energy generation capacity added due to CDMs worldwide since 2005 – it has been observed that approximately 10% of this global installed capacity came from developing countries which are non-Annex I Parties contributing significantly towards climate change actions globally (OECD 2016). Therefore, it is evident that when properly implemented with stringent guidelines in place -CDM projects not only serve their purpose in terms of mitigating effects associated with climate change but also bring about various socio-economic benefits locally at community level – all while ensuring environmental integrity along the way. Figure 5.6 summarizes the CDM.

To ensure the requirements for CDM projects are met, an approval process is conducted before they can start generating carbon credits to be sold on international markets (Blok, 2018; Brandenberger et al., 2022; Laycock & Mitchell, 2019; Sattler & Schroter, 2022; Van Popering-Verkerk et al., 2022). This process ensures that all CDM projects meet established criteria and standards of quality prior to their implementation (Dietz & Zhang, 2014).

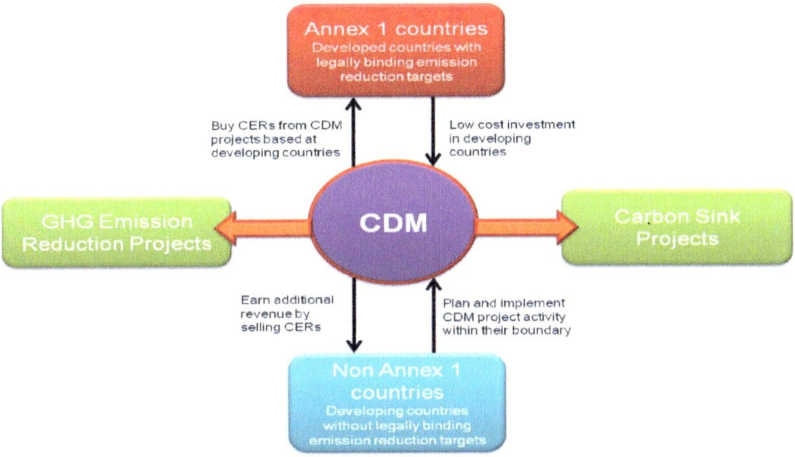

Figure 5.6. An overview of the clean development mechanism (CDM).

Source: Green Clean Guide.

In addition, this approval process allows governments to track each project's progress in order to verify that it meets its stated goals (Gong et al., 2019). Subsequently, the World Bank estimates that up to 40% of potential CDMs may fail due diligence if not assessed through a rigorous pre-approval procedure (Kammen & O'Connell, 2011). Moreover, research has shown that approved CDMs produce more successful outcomes than those without such oversight – with avoided emissions increasing by almost 50% when compared against unapproved counterparts (Verbruggen et al., 2016). Thus, having an effective mechanism in place provides assurance both domestically and internationally regarding the efficacy of these initiatives.

Once approved, Certified Emission Reduction (CER) credits from CDM projects may be traded on international markets (Bailey et al., 2012; Bjerkan & Ryghaug, 2021; Fried et al., 2022; Niles et al., 2021; Willems & Giezen, 2022). According to a 2018 study by the International Institute for Sustainable Development (IISD), "the overall market value of CDM CERs transacted in 2017 was estimated at over US$500 million" (Peters & Corfee-Morlot, 2018). This amount reflects an increase since 2016 and is indicative of growing global interest in carbon trading as well as increasing demand for emission reduction units such as those generated through CDMs. Due to supply and demand forces within these markets, prices vary greatly depending on factors like regulatory changes or geopolitical uncertainty that influence investor confidence levels. For example, Peters and Corfee-Morlot note that although the price per credit had decreased slightly between 2013 and 2014 due to political transition in some countries working with CDMs – it still remained higher than 2008 averages when compared to other types of emissions reductions schemes (2018). These fluctuating prices are not only beneficial for project developers but also demonstrate how environmental regulations can affect economic activity which serves as a reminder that climate action must take into account both policy concerns while considering financial incentives necessary for implementation.

In summary, the CDM is an international agreement that was established under the Kyoto Protocol in 1997. It allows developed countries to invest in projects that reduce emissions of greenhouse gasses (GHGs) in developing countries, and receive credits for doing so. These credits can then be used by the investing country to meet its own emission reduction targets. The CDM has been linked to the Paris Agreement, which sets out a global framework for reducing GHG emissions and limiting climate change. Under this agreement, all participating countries are required to take action towards achieving their NDCs. The CDM provides a mechanism through which these NDCs can be achieved more quickly and cost-effectively than other methods. By providing financial incentives for investments into low-carbon technologies and renewable energy sources, it helps accelerate progress towards meeting NDCs while also creating economic opportunities for both investors and host communities.

5.4. THE PARIS AGREEMENT

The Paris Agreement is an international agreement that was adopted in 2015 to combat climate change (Hale & Roger, 2014; Hoffmann & Villamayor-Tomas, 2023; Krause et al., 2021; Tuda & Machumu, 2019). It sets out a framework for countries to reduce their greenhouse gas emissions and limit the global temperature increase to below 2 degrees Celsius above pre-industrial levels by 2100. The Agreement also encourages countries to use carbon markets and emissions trading as part of their efforts to meet these goals. Carbon markets are economic instruments that allow governments, businesses, or individuals to buy and sell allowances for emitting certain amounts of greenhouse gasses into the atmosphere. These markets create incentives for reducing emissions by allowing those who emit less than their allocated amount of pollution credits (or "allowances") to sell them on the market at a profit. Emissions trading systems are similar but involve two different types of participants: buyers and sellers who trade emission permits with each other rather than buying from or selling directly back into the government system. Under the Paris Agreement, participating countries have committed themselves towards setting up national carbon pricing mechanisms such as taxes, cap-and-trade systems, or voluntary agreements between companies which will help them reach their targets set under this agreement more efficiently while creating additional revenue streams through auctioning off unused allowances within these schemes. This has led many experts in environmental economics to believe that carbon markets could be one of most effective tools available when it comes to tackling climate change globally over time if implemented correctly across all nations involved in this agreement. Figure 5.7 shows that by the 21st of April 2021, most territories across the globe had either signed or ratified the Paris Agreement. This represents broad-based acceptance of many of its premises.

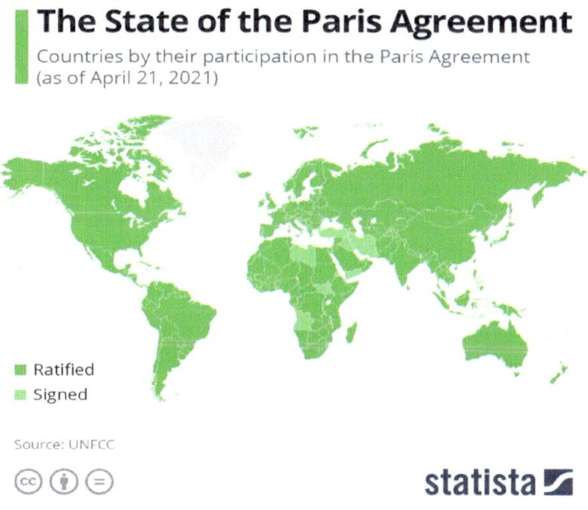

Figure 5.7. The state of the Paris agreement by 2021.

Source: Statista.

The Paris Agreement established a framework for collective climate action that serves as an important foundation to reduce global greenhouse gas (GHG) emissions (Bodin, 2017; Lubell & Robbins, 2022; National Intelligence Council, 2017; Sattler & Schroter, 2022). The agreement outlines the need to limit warming of the planet's surface temperature to 2°C and strive for 1.5°C, with countries committing to their own targets in NDCs. One way this has been achieved is through carbon markets and emissions trading systems, which offer economic incentives for reducing

GHG pollution by allowing nations or organizations to buy or sell emission credits among themselves. For example, according to Shrestha et al., "In line with the NDCs under the Paris Agreement on Climate Change 2015, China launched its national ETS [Emissions Trading Scheme] covering the power generation sector at the end of 2017" (2019, p. 844). This type of system puts limitations on some polluters while rewarding those who can reduce their emissions within a given time frame; thus providing powerful financial incentives towards environmental protection goals set out by the Paris Agreement (Shrestha et al., 2019). Accordingly, the World Bank estimated that if all 20 pilot programs—which included California Cap-and-Trade System—were fully implemented across three sectors: electricity generation, industry and transportations then total annual CO_2e reductions would be over 4 billion tons per year globally from 2020–2030 compared against business as usual scenarios without these policies in place (World Bank Group 2018). These numbers are promising evidence that further implementation of such market systems could have significant positive impacts when it comes meeting our commitments outlined in The Paris Agreement.

The agreement set out an ambitious goal to keep the average global temperature "well below 2°C above pre-industrial levels and pursue efforts to limit the temperature increase even further to 1.5°C" (United Nations Framework Convention on Climate Change [UNFCCC], 2016). This target is important for limiting some of the most severe impacts of climate change such as extreme weather events, decreased water availability, rising sea levels, and species extinction (European Commission Directorate General Environment & Oceans & Fisheries [EC DGOEF], 2019). Recent studies have found that meeting this two degree Celsius target would require significant reductions in greenhouse gas emissions compared with current emission pathways or business-as-usual scenarios (Höhne et al., 2017; Van Vuuren et al., 2018). For example, Höhne et al.'s study estimated that reducing emissions from 41 gigatons per year to 25 gigatons per year between 2020 and 2030 is necessary for keeping warming below two degrees Celsius over the long term (2017). Similarly, Van Vuuren et al.'s study predicted a peak CO_2 concentration of 450 parts per million before 2050 if we are able stay within this two degree target according to their modeled pathways based on various economic factors like population growth and energy demand patterns (2018). Limiting warming even further than two degrees Celsius presents much more challenging goals. Achieving a one point five-degree Celsius maximum rise will likely require dramatically steeper cuts in GHG emissions as well as large scale removal processes known as carbon dioxide removal technologies which actively remove carbon dioxide from the atmosphere through direct air capture methods or biological sequestration techniques such afforestation initiatives (Solomon et al., 2009; Smith et al., 2014). While these interventions can play a role in mitigating climate change they still remain largely unproven technology thus far due primarily to cost constraints (Smith, 2016). Additionally, many countries may be unwilling to engage in costly endeavors given limited resources when other pressing priorities arise domestically (Figure 5.8 for an evaluation of the performance of different countries against the Paris Agreement). Therefore, achieving true success requires international cooperation among nations both economically developed and developing coupled with a strong commitment adhere policy objectives outlined Paris Agreement reduce overall worldwide carbon footprint significantly enough reach desired temperature targets.

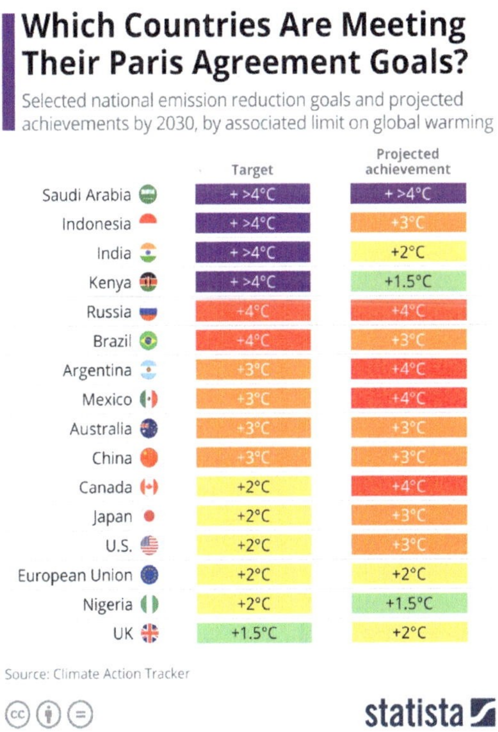

Figure 5.8. Evaluating the performance of countries under the Paris Agreement.

Source: Statista.

Under the Agreement, all countries agreed to take ambitious efforts to combat climate change by reducing their net greenhouse gas emissions through NDCs with an aim of achieving a balance between sources and sinks of these gasses in the second half of this century (Koefoed, 2019; Laycock & Mitchell, 2019; Leventon et al., 2017; Partelow et al., 2020; Sanders et al., 2017). The NDCs are legally binding commitments that set out actions each country will take over time towards meeting its commitment to keep temperatures from rising more than 2 degrees Celsius above pre-industrial levels. In order to achieve this goal, countries committed themselves to taking specific steps such as setting economy-wide emission reduction targets for 2030 or earlier; developing long term strategies for low greenhouse gas emissions development pathways leading up until 2050; investing resources into renewable energy production capacity; improving access to clean energy services; and increasing adaptation measures such as early warning systems against extreme weather events. In addition, developed countries also pledged financial support – totaling around US$100 billion per year – along with technology transfer and technical assistance needed by developing nations so they can adapt effectively while transitioning away from high carbon economies into greener models of development. This includes both direct transfers from public funds managed by multilateral institutions like the Global Environment Facility (GEF) but also private sector investments facilitated through green finance instruments like climate bonds or green loans which reward companies making positive environmental impacts though their operations or products/services offered.

This collective effort has already seen some success (Daher et al., 2019; Orach & Schluter, 2016; Rozance et al., 2019; Therville et al., 2019; Vasconcelos et al., 2015): since 2005 there have been record reductions in CO_2 equivalent global emissions partly due largely due increased investment made available after signing onto the Paris Agreement (PA). According to 2020 estimates published by UNEP's Emissions Gap Report, global GHG decreased 1% compared with 2019 output representing an estimated 4 GtCOe avoided emission when compared with 2010 levels. However, despite these improvements further ambition is necessary if we want to reach our ultimate objective which is limiting warming below 1.5°C according to IPCC reports released in 2018. To do this, several studies suggest deeper cuts must be achieved quickly: at least 7 Gigatons annually during 2021–2030 period if we are going to meet Net Zero target no later than mid-century (IPCC Special Report on Global Warming). It is clear then that much work still needs done before we get close to reaching our goals outlined within the PA framework but it provides us with foundations upon which to build future generations resilient societies capable of mitigating the worst effects associated with human induced climate change.

Under the Agreement, all countries agreed to make ambitious efforts to combat climate change by reducing their net greenhouse gas emissions through NDCs with an aim of achieving a balance between sources and sinks of these gasses in the second half of this century (van Buuren & Eshuis, 2009; Van Popering-Verkerk et al., 2022; Vantaggiato et al., 2023; Vasconcelos et al., 2015). The Paris Agreement of 2015 exemplifies a historic global commitment to combatting climate change, compelling all participating countries to pledge ambitious efforts to reduce their net greenhouse gas emissions through Nationally Determined Contributions (NDCs) with a calculated aim of achieving a balance between sources and sinks of these gases in the second half of this century. A premier case study illustrating this commitment can be found in the exemplary endeavors undertaken by Denmark, a country that has taken the shared responsibility outlined in the Paris Agreement to considerable heights. Denmark's Climate Act of 2020 sets the stage for this nation's policy landscape, obligating it to achieve 70-percent reduction of its 1990 greenhouse gas emissions by 2030, and to become climate-neutral, reaching net-zero emissions by 2050. This is significantly more ambitious than their initial NDC submitted under the Paris Agreement. Integral to Denmark's approach is the employment of an annual climate action plan incorporating new concrete initiatives based on the most recent climate prognoses to meet these targets. Contributing to the networked climate governance strategy, Denmark actively engages in dialogues with industry stakeholders, local municipalities, and citizen engagement initiatives to ensure extensive consensus-building and transparency. The Danish Council on Climate Change, an independent body of climate experts, monitors the nation's progress, rendering a yearly status to adjust policies and strategies accordingly, a model reflective of an adaptive management framework. Denmark's robust strategy displays its commitment to the Paris Agreement and exemplifies a forward-facing, action-oriented approach to mitigating climate change impacts. Through the meticulous integration of governance, expert monitoring, and citizen involvement, Denmark presents a template of best practices for countries striving to achieve ambitious climate goals while maintaining a balance of greenhouse gas sources and sinks.

This collective effort has already seen some success: since 2005 there have been record reductions in CO_2 equivalent global emissions partly due largely due increased investment made available after signing onto the Paris Agreement (PA). The collective endeavor towards reducing global greenhouse gas emissions, galvanized by the 2015 Paris Agreement, has seen notable progress: emissions of CO2 equivalent have experienced record reductions since 2005. A salient beneficiary of this international climate action commitment, as illustrated through the case study of the United Kingdom, has benefitted from investments channeled towards green initiatives post-Paris Agreement. The United Kingdom, renowned for its rigorous climate policies, has witnessed an impressive 38% reduction in emissions since 1990, the backbone of this reduction being a record drop in carbon dioxide emissions since 2005. The sustained investment into wind, solar, and nuclear energy sources, rendered possible by the enhanced financial resource mobilization precipitated by the Paris Agreement, attributed considerably to such reductions.

Critically, the UK's Carbon Budgets, initiated by the 2008 Climate Change Act, steered the policy environment towards structured emissions cutting, setting statutory limits on the amount of greenhouse gases the UK can emit over a 5-year period. The Carbon Budgets reorient the nation towards low-carbon, climate-resilient growth, attracting investments in renewable energy, electric vehicle technology, and energy efficiency. Furthermore, emissions cuts have been achieved while the UK economy has simultaneously grown, defying the presumed trade-off between economic prosperity and reduced emissions. The UK's experience post-Paris Agreement, therefore, underscores the potential of industrialized nations in realizing substantial reductions in CO2 equivalent emissions, attributing the success largely to the increased investments in sustainable alternatives, which have driven a transition away from a carbon-based economy. Through such demonstrative progress, the UK substantiates how harmonizing economic growth with environmental conservation is a feasible and sustainable path forward.

Finally, the agreement created an international cooperative mechanism, known as the international emissions trading system (IETS), which allows developed countries to purchase emission reduction credits from developing countries that have achieved more than their required reductions targets (Cody, 2018; Dey et al., 2017; Mar et al., 2023; Watabe & Gilby, 2020; Wyborn, 2015; Yates, 2012). This provides a financial incentive for investment in low-carbon technologies and increases access to capital for poorer nations who are not able to finance such projects themselves. Literature has demonstrated how this can be beneficial both economically and environmentally. For example, research by Stadelmann et al., 2017 found that IETS could reduce global abatement costs whilst also providing additional economic benefits through technology transfer or foreign direct investments into recipient countries. Furthermore, it was estimated by Wang et al., 2018 that over $1 trillion would need to be invested annually in clean energy infrastructure between 2015–2050 if the world is serious about meeting its shared climate goals set out within the Paris Agreement. Therefore, these mechanisms provide an effective way of incentivizing those investments needed to transition away from fossil fuels towards cleaner alternatives at scale globally – helping ensure no nation gets left behind when transitioning toward lower carbon economies and societies worldwide. Furthermore, recent statistics published by International Emissions Trading Association further back up these claims: according to them in 2019 alone around 400 million tons of CO_2e were traded under various international cooperative mechanisms established under Article 6 of the Paris agreement – with China accounting for almost 50% of all transactions made via these markets during this period. These figures highlight just how important these types of initiatives have become on a macro level; clearly demonstrating they're already beginning making a real difference in terms of tackling climate change effectively on a global scale.

In summary, the Paris Agreement is an international agreement that was adopted in 2015 to address the global challenge of climate change. The main goal of the agreement is to keep global temperature rise this century well below 2 degrees Celsius above pre-industrial levels and pursue efforts to limit it even further, to 1.5 degrees Celsius.

To achieve this goal, countries have committed to reduce their emissions by setting NDCs. In addition, the Paris Agreement also aims at increasing economic efficiency through a number of measures such as promoting sustainable development and investments in renewable energy sources; enhancing adaptive capacity and resilience; strengthening cooperation between developed and developing countries on technology transfer; providing financial resources for adaptation activities; creating incentives for reducing greenhouse gas emissions from deforestation and forest degradation (REDD+); improving transparency systems related to mitigation actions taken by Parties; etc. These measures are expected not only help reduce carbon dioxide emissions but also create new jobs, increase productivity, stimulate innovation, promote trade opportunities among nations while helping them adapt better with changing climatic conditions.

5.5. ECONOMIC EFFICIENCY

Economic efficiency under carbon markets and emissions trading is a topic that has been gaining traction in recent years (Buchtmann et al., 2022; Mar et al., 2023; Perz et al., 2022; Therville et al., 2019; Tosun & Schoenefeld, 2017). This is due to the fact that these systems have become increasingly popular as a way of reducing greenhouse gas emissions while also providing economic incentives for businesses to reduce their environmental impact. Carbon markets and emissions trading are based on the idea of creating an economic incentive for companies to reduce their emissions by setting up a market where they can buy or sell permits allowing them to emit certain amounts of greenhouse gasses. Companies must purchase additional permits if they exceed their allotted amount, thus incentivizing them to reduce their emissions in order to save money. In addition, this system allows governments and other entities (such as non-profits) who wish to invest in green initiatives access capital from those buying emission credits at higher prices than what was initially paid out when purchasing the permit. The goal of such systems is twofold: firstly, it provides an effective means for reducing overall global greenhouse gas levels; secondly, it creates an efficient market mechanism which encourages investment into clean energy technologies and projects without relying solely on government subsidies or taxes. Ultimately, this leads towards greater economic efficiency through increased competition between firms within the industry leading towards lower costs associated with production processes involving fewer pollutants being released into our atmosphere. Figure 5.9 shows this type of leadership in urban settings.

Allocation efficiency is an important economic consideration when it comes to collective climate action and networked climate governance (Atalan Helicke, 2019; Brewer, 2013; Fisher & Nasrin, 2021; Olivier & Berardo, 2022; van Bers et al., 2019). It refers to the ability of a system or policy to distribute resources in a manner that maximizes their utility, cost-effectiveness, and overall value within society. Research has indicated that allocation efficiency can be improved through increased participation in decision-making processes related to collective climate action (Vörösmarty et al., 2020). This includes allowing for more equitable distribution of financial aid by providing support for vulnerable countries and communities who are disproportionately impacted by the effects of climate change (Adger et al., 2012). Additionally, increasing transparency around resource allocations increases trust among stakeholders which helps ensure effective implementation of strategies designed towards achieving emissions reductions goals (Lemos & Rocha, 2011). Studies have also suggested using market mechanisms such as carbon taxes as incentives for reducing greenhouse gas emissions could lead to greater allocative efficiencies than traditional command-and-control policies like quotas on CO_2 emission levels (Connor & Anderson 2019).

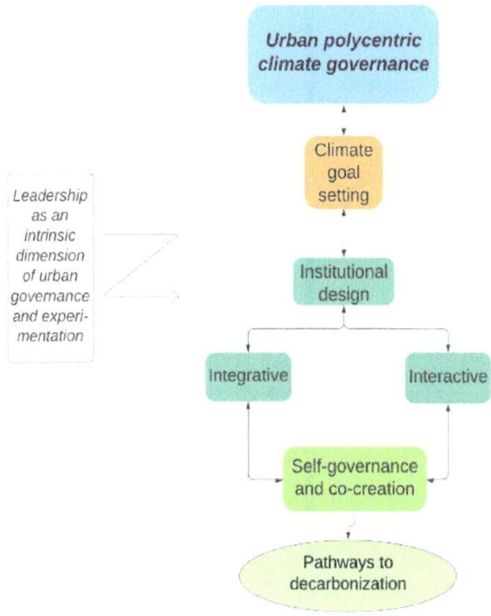

Figure 5.9. Leadership of polycentric urban climate governance.

Source: Vedeld et al. (2021).

An example of how allocation efficiency was implemented successfully is seen from China's 'Circular Economy Law' which sought to reduce environmental pollution while promoting economic growth through efficient use of resources. The law incentivized enterprises and households with tax benefits who adopted eco-friendly practices such as energy conservation measures thus leading them towards better performance outcomes over time without compromising effectiveness due costs incurred during implementation (Raboinyana et al., 2009; Liu 2017). A study conducted looking at rural water supply systems found that implementing pro-poor tariff structure helped lower income individuals access clean drinking water while simultaneously ensuring optimal utilization rates amongst users across different socio-economic sectors (Jiao 2014); this reveals the importance placed on creating inclusive solutions which prioritize equality rather than just focusing on efficacy alone.

An understanding of allocation efficiency can help inform decisions about cost effectiveness when dealing with collective climate actions since they both deal with making sure resources are being used economically in order to achieve desired objectives efficiently (Atalan Helicke, 2019; Bailey et al., 2012; Bernstein & Hoffmann, 2018). Allocating funds appropriately ensures maximum return on investment thereby helping organizations maintain budget constraints while still remaining productive; however cost effectiveness focuses more specifically on minimizing expenditure associated with producing goods or services needed for operations. Both concepts should be considered together when developing strategies aimed at solving complex issues related to global warming so we make sure our efforts yield long term sustainable results instead short-lived gains based solely on numerical values calculated from current data sets available but do not fully account for external factors influencing progress.

Climate change is one of the most pressing global issues of our time, with far-reaching implications for economic activity and human well-being (Brewer et al., 2017; Dal Poz et al., 2022; Diop & Remvikos, 2016; Laycock & Mitchell, 2019; Watabe & Gilby, 2020). The need to take collective action towards mitigating its effects has been widely recognized in recent years, leading to the development of numerous initiatives aimed at promoting climate governance and reducing emissions. In this context, cost-effectiveness plays an important role as it helps determine which strategies are best suited for achieving a given goal efficiently while minimizing overall costs. This is particularly important given the patterns of global climate change expenditure (Figure 5.10).

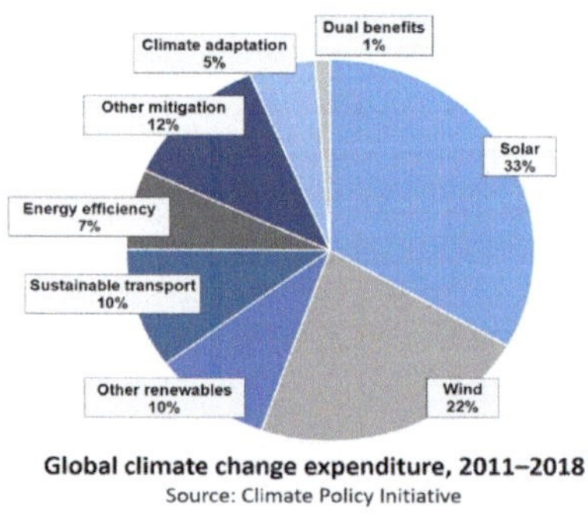

Figure 5.10. Global climate change expenditure between 2011 and 2018.

Source: Climate Policy Initiative as quoted by *ÓhAiseadha* et al. (2020).

The concept of cost-effectiveness can be applied to all forms of collective climate action, from international agreements such as the Paris Agreement to local or regional efforts like carbon taxes or cap-and-trade schemes (Bodin, 2017; Chaudhury et al., 2017; Morrison et al., 2017; Rahimi-Feyzabad et al., 2022; Virah-Sawmy et al., 2019). For example, research suggests that implementing energy efficiency standards could reduce greenhouse gas emissions by up to 30–50% compared to business as usual levels (Rees et al., 2020). This kind of strategy offers a high level of effectiveness relative to its cost since technological improvements typically require relatively low upfront investments but offer substantial long term returns in terms of reduced emissions and increased resource efficiency. Similarly, networked approaches have also proven successful in helping countries coordinate their actions on climate change more effectively (Fazey et al., 2014). Through networks like Climate Action Network International (CAN) or Global Covenant Cities Initiative (GCCI), governments can pool resources and share knowledge about how best tackle challenges related to mitigation and adaptation measures without incurring significant additional expenses. Such distributed models not only help increase efficiency but also foster collaboration between different actors involved in tackling climate change locally or globally—allowing them better leverage their respective strengths while avoiding unnecessary duplication in effort or spending.

Cost benefits associated with taking effective collective action on climate should not however overshadow other considerations related to price stability when making policy decisions (Bernstein & Hoffmann, 2018; Luthe & Wyss, 2016; Salmoral et al., 2020; Sanders et al., 2017; van Buuren & Eshuis, 2009). While policies aiming at cutting down emissions may lead initially to higher prices due to government interventions, such incentives will often result over time into cheaper goods through innovation spurred by regulations forcing firms to adopt cleaner solutions (Burton & Wesselbaum 2019). Therefore, being mindful of both short term impacts alongside longer term gains achieved by establishing adequate market conditions that promote beneficial investment flows is essential when trying to balance environmental concerns against economic growth objectives. Climate change has become a pressing problem for the entire world, with economic efficiency being heavily impacted by collective climate action. To better understand this issue, it is important to assess how price stability plays into the equation of economic efficiency and networked climate governance.

Research suggests that when governments implement policies designed to reduce greenhouse gas emissions, these measures can also increase market liquidity which helps enhance overall price stability (Schneider et al., 2018). This increased liquidity allows more players to enter financial markets while reducing transaction costs associated with trading assets such as stocks or bonds (Mankiw & Taylor 2019). Therefore, stable prices are essential components of both economic efficiency and collective climate action due to their ability to facilitate better investment decisions within global markets. Furthermore, research indicates that networked climate governance structures have been particularly successful in promoting not only environmental protection but also efficient pricing mechanisms across countries (Duitsman et al., 2016). These networks allow countries from all over the globe to collaborate on various projects related to energy security and climate change mitigation efforts through shared data sources and policy instruments such as carbon taxes or renewable subsidies (Eriksson et al., 2017). By creating an interconnected system of incentives between individual nations aimed at achieving greater sustainability goals collectively, these networks incentivize participants towards making investments geared towards improving long-term returns on capital rather than short-term speculative gains. As a result, they help create conditions where asset prices remain relatively stable regardless of external shocks occurring outside specific economies or regional blocs – ultimately leading towards lasting improvements in terms of economic efficiency worldwide.

Price stability is an integral factor influencing economic performance linked closely with both collective climate actions as well as networked forms of international governance arrangements (Anderton & Setzer, 2018; Granderson, 2017; Hawkins et al., 2016; Kithiia, 2015; van Bers et al., 2019). When applied effectively through government initiatives like carbon taxes or renewable subsidies along with cooperation amongst participating nations via multi level platforms; market liquidity increases allowing investors access lower cost transactions while still maintaining steady levels for various asset classes – thereby helping strengthen macroeconomic outcomes globally even amidst times uncertainty regarding future trends around resource utilization patterns resulting from our changing environment. Market liquidity is an important factor in the economic efficiency of collective climate action and networked climate governance. Market liquidity refers to how easily assets can be bought or sold without affecting their market price significantly, as well as the amount of volume that trades on a given market daily. Market liquidity is essential for efficient markets because it allows investors to quickly enter and exit positions with minimal cost, which helps ensure financial stability and encourages investment and trade. It also has implications for trust between actors within any trading system – if participants cannot freely buy or sell assets then they may not have confidence in the system's reliability. Literature has long highlighted the importance of market liquidity when discussing economic efficiency under collective climate action (Tosun & Schoenefeld, 2017; Tuda & Machumu, 2019; van Bers et al., 2019; van Buuren & Eshuis, 2009). For example, one study found that high levels of transaction costs (such as those associated with low market liquidity) are likely to reduce investor participation in emissions trading schemes such as cap-and-trade systems (McKinsey & Company 2006). This could lead to reduced demand from buyers who view transaction costs as too great, resulting in lower prices for carbon credits than would otherwise occur with more liquid markets (OECD 2011). A lack of reliable pricing information due to limited volumes traded will also make it difficult for governments and other stakeholders involved in these schemes to plan appropriately (Yamauchi et al 2018). Likewise, statistics demonstrate that higher levels of market liquidity are strongly correlated with increased economic benefits derived from emissions trading activities related to global warming mitigation efforts (Newell et al 2012). One analysis estimated that increasing average daily turnover by just 5% results in up to 4% greater aggregate welfare gains per year compared with less liquid markets (Stavins 2013). Clearly this indicates substantial potential improvement through better management practices aimed at boosting overall transaction speeds and deepening existing markets.

Transparency and accountability play major roles alongside improved market liquidity when considering effective approaches towards achieving maximum economic benefit from collective climate action initiatives worldwide (Hamilton & Lubell, 2018; Harmon et al., 2023; Hawkins et al., 2018). Increased transparency around emission reductions achieved by each participant along with regular reporting requirements facilitates oversight; while stronger enforcement measures against noncompliance help maintain public trust by ensuring consistency across all players regardless of size or location (Stern 2017). Such measures become even more critical when taking into account broader issues like social equity concerns surrounding environmental protection policies implemented globally (Gorski 2019). Ultimately, both transparency/accountability factors combined provide safeguards necessary for keeping various parties accountable during implementation phases while allowing them access needed resources so they can effectively meet obligations outlined within relevant frameworks (UNFCCC 2020). Therefore, it is clear that improving overall levels of market liquidity plays a key role promoting successful outcomes via enhanced economic efficiency through collective climate actions governed via networked structures. Consideration must however always be given simultaneously regarding additional dimensions such privacy rights, data security etc., to gain full understanding about the effectiveness degree attainable via specific approach taken over the time period concerned (Konig& Dalhuisen 2015).

Transparency and accountability are essential components of economic efficiency under collective climate action and networked climate governance (Gregg et al., 2023; Lee & Koski, 2014; Morrison et al., 2017; Ostrom, 2014; Panditharatne, 2016; Sara et al., 2016). While transparency primarily refers to the public disclosure of information, accountability implies a much broader concept in which individuals or organizations are held responsible for their actions. Both factors play an important role in creating effective measures that reduce emissions and ensure equitable global access to resources. Literature has identified several key areas where transparency helps promote efficient economic decision-making related to tackling climate change, one example being carbon pricing policies (Chen et al., 2020). Carbon pricing systems can be made more transparent by regularly releasing data on companies' compliance with emission reduction targets as well as corresponding penalties imposed when these standards are not met (Tollefsen & Jagannathan, 2016). This heightened level of visibility incentivizes businesses to invest in low-carbon technology while at the same time providing oversight from regulatory agencies charged with ensuring rules pertaining to this policy area are followed. Additionally, increased transparency may foster greater cooperation between actors across different sectors such as renewable energy producers and fossil fuel suppliers due to improved trust through reliable communication (Lee et al., 2021).

Accountability is also necessary for achieving successful outcomes in regards to addressing the effects of climate change (Kim et al., 2019; Kithiia, 2015; Koefoed, 2019). For instance, research indicates that if governments have clearly outlined strategies concerning how they will uphold commitments established by international agreements like the Paris Agreement then it increases countries' likelihood of meeting those goals over time (Rasmussen & Olesen, 2018). Furthermore, local initiatives often require citizens' engagement so there must be some sort of mechanism whereby people feel accountable for their individual contributions towards reducing greenhouse gas pollution levels (Fernández de la Peña et al., 2019; Vandenbosch et al., 2017). These examples demonstrate why both transparency and accountability should remain integral parts within any strategy designed around collective efforts against global warming since they create incentives among all stakeholders involved while promoting collaboration at multiple scales simultaneously.

The combination of higher levels of both transparency and accountability leads directly into systemic changes needed for society transition away from reliance upon traditional forms of energy sources such as coal towards newer cleaner alternatives like solar power generation plants. Such transformations involve major shifts away from existing institutions governing resource allocation practices thereby necessitating far reaching reforms throughout many aspects of our societies including political structures plus cultural norms associated with consumption habits (Shroder & Van Der Geer Papezováa, 2018). Consequently implementations intended to increase participation amongst wider audiences requires considerable investment in terms of financial resources along open forums allowing civil society movements leverage influence to shape future policy decisions align interests affected populations concerned about environment sustainability issues (Qorri & Baldacchino, 2019). As a result creating a framework wherein each actor exercises responsibility implementing suitable solutions vital components ensuring lasting impacts resulting sustainable development objectives set forth various regulations adopted internationally significantly reduces risks failure long term success programs initiated tackle pressing environmental crises facing the world today (Garcia Moreno et al., 2020). Overall Transparency enhanced Accountability serve crucial roles mitigating harmful effects caused human activities planet earth efficiently economically viable manner. Thus, incorporating elements both expected lead tangible gains ameliorating repercussions arising Climate Change consequently having positive implications entire living ecosystems interconnectedness work together achieve common goal preserving fragile ecosystem we inhabit intergenerationally secure bright future generations come.

CHAPTER 5 SUMMARY

In conclusion, carbon markets and emissions trading are tools that can help to reduce global greenhouse gas emissions. They provide incentives for companies to reduce their own emissions as well as providing financial support for those who do so. Carbon markets also create a market-based framework in which the cost of reducing emissions is shared fairly among all involved parties. While these mechanisms have been successful in some cases, they cannot be relied upon to address climate change alone; collective action by governments and other stakeholders must play an important role if effective mitigation efforts are to be achieved. It has become increasingly apparent that more comprehensive approaches will be needed if we want to ensure lasting success in combating climate change on a large scale. Systemic changes such as shifting our energy systems away from fossil fuels or transitioning towards sustainable agricultural practices may prove necessary if real progress is ever going to be made against this issue. This chapter has discussed how carbon markets and emission trading can help us take steps forward, but ultimately systemic approaches must come into play if transformative results are desired—and it is up to everyone else working within the field of networked climate governance to work together with one another toward making this happen. The next chapter "Systemic Change" will explore further what types of systemic solutions should now be put forth at both local and international levels, while examining existing initiatives already underway that demonstrate just how powerful concerted action between multiple actors can truly be when tackling complex problems like climate change head on.

CHAPTER 6

Elements of Systemic Change in Climate Action

CONTENTS

The sixth chapter of the book Collective Climate Action and Networked Climate Governance, titled "Systemic Change," examines how collective climate action can foster systemic change. It looks at the concept of systemic transformation from a theoretical perspective to explore what it means in practice for tackling climate change. The chapter begins by discussing why conventional approaches have failed to address global warming sufficiently and explains why shared leadership is needed instead. It then moves on to discuss different ways that collective climate actions can help facilitate systemic change through new ideas, networks, alliances and organizations. Finally, the chapter reviews potential pathways for achieving transformative outcomes based on these principles with a focus on scaling up local initiatives into wider systems changes that will ensure greater success in addressing climate challenges globally. At its core, this chapter argues that without understanding the dynamics of system-level governance we are unlikely to make meaningful progress towards mitigating or adapting successfully to our changing environment; therefore efforts should be directed toward creating collaborative cross-sectoral strategies which recognize multiple actors as well as their respective roles within larger structures such as markets and societies. We must also strive for an inclusive approach centered around community engagement where those most directly affected by climatic shifts get involved in strategic conversations about long term solutions – only then can equitable decisions be made regarding policy interventions and resource allocations essential for implementing successful adaptation plans over time. In addition, innovative technology platforms need to become more widely utilized if they are going to support both public sector goals aimed at reducing emissions while simultaneously helping private enterprises overcome financial barriers related to capital investments that are necessary to transition away from fossil fuels (e.g., renewable energy sources). Ultimately, any strategy designed to create broad scale transformational impacts needs to pay close attention to existing power relations between various stakeholders to maximize positive results at all levels of society including individual households right up to international institutions like UNFCCC COP meetings.

6.1. THE UNITED NATIONS FRAMEWORK CONVENTION ON CLIMATE CHANGE (UNFCCC)

The United Nations Framework Convention on Climate Change (UNFCCC) is an international treaty that was adopted in 1992 and entered into force in

1994. The UNFCCC sets out a framework for countries to work together to reduce greenhouse gas emissions, mitigate the effects of climate change, and adapt to its impacts (Currarini et al., 2016; Hale & Roger, 2014; Neville et al., 2019; Teodoro et al., 2021; Vantaggiato et al., 2023). The UNFCCC has been instrumental in driving systemic change by providing a platform for governments around the world to come together and agree on measures that will help address global warming (Figure 6.1). It also provides guidance on how countries can implement policies related to climate change mitigation and adaptation. Additionally, it encourages collaboration between different stakeholders such as businesses, civil society organizations, local authorities, indigenous peoples' groups etc., which helps ensure that all voices are heard when making decisions about climate action. Overall, the UNFCCC serves as an important source of information about current trends in global warming and potential solutions for addressing them. Through regular reporting requirements from signatory nations under the convention's Kyoto Protocols agreement, data is collected regarding national efforts towards reducing emissions or adapting to changing climates – allowing us greater insight into what works best at tackling this pressing issue worldwide.

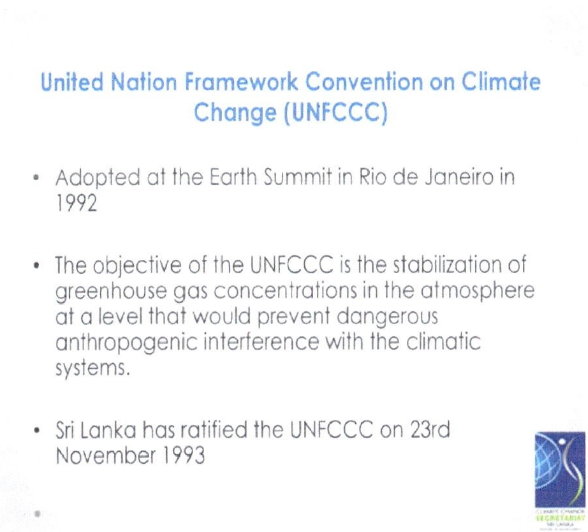

Figure 6.1. The United Nations framework convention on climate change (UNFCCC).

Source: Alexis Cameron via SlidePlayer.

The five principal aspects of The United Nations Framework in Systemic Change under collective climate action and governance are as follows:

- Strengthening the global response to climate change by enhancing international cooperation, including through multilateral environmental agreements;
- Enhancing national efforts for effective implementation of commitments made at the Paris Agreement on Climate Change;
- Promoting sustainable development pathways that reduce emissions while increasing resilience to climate impacts;
- Supporting countries in their transition towards low-emission and resilient societies; and
- Facilitating access to finance, technology transfer, capacity building and other forms of support for developing countries' actions on adaptation and mitigation. We will now explore these aspects in more detail:

Climate change is one of the most pressing issues humanity faces today. The UNFCCC was established in 1992 to address climate change and its effects by strengthening international cooperation between nations, including through multilateral environmental agreements (MEAs). MEAs are a type of international agreement that involve multiple countries or organizations working together towards common goals related to the environment (Hudson & Boergers, 2018). The UNFCCC has been successful in fostering global response to climate change due to increased awareness and participation from various stakeholders such as governments, intergovernmental organizations, non-governmental organizations, civil society actors, businesses, researchers and other interested parties. As of 2020 there are 197 Parties signed onto this Agreement which represents almost all countries worldwide ("Parties to the convention" n.d.). This widespread support for tackling climate change demonstrates how effective MEAs can be at encouraging international collaboration on complex environmental challenges. In addition to increasing awareness about climate change globally and promoting collective action among signatory states; standards set out under MEAs can also help ensure conformity with agreed upon principles amongst participating nations. For example, research conducted by Kallbekken et al., (2018) found that compliance with emissions targets set out under the Kyoto Protocol improved significantly when monetary penalties were imposed for violations of protocol rules compared without any sanctions being applied against violators. Their results suggest that establishing enforceable regulations within MEAs helps increase adherence rates especially when appropriate consequences are implemented accordingly against those who fail to comply with agreed frameworks. While enforcement mechanisms alone will not solve problems posed by climate change, it does demonstrate benefits associated with strengthening international cooperation around shared objectives via collaborative arrangements such as an MEA framework like UNFCCC.

The Paris Agreement on Climate Change under the UNFCCC was adopted in 2015 and came into force in 2016. It is a legally binding international agreement that seeks to limit the global temperature increase within 2 degrees Celsius above pre-industrial levels, with an aim of keeping it below 1.5 degrees Celsius by 2100 (Kumar & Gupta, 2018). The target of limiting emissions has been set across all countries as well as sectors such as energy production, land use and forestry management (UCSUSA, 2020). In order for these commitments to be effectively implemented at a national level, there needs to be increased effort from governments around the world to create effective policies and take actionable steps towards achieving climate targets (Kelman & Clark-Ginsberg, 2022; Kim et al., 2019; Kithiia, 2015; Koefoed, 2019). One way of doing this is through introducing carbon pricing policies which have seen success when implemented properly. Carbon pricing puts a price tag on carbon dioxide or other greenhouse gas emission-based activities so that they become more expensive relative to alternatives that are less polluting like renewable sources of energy (Ravikumar et al., 2019). This encourages companies and individuals alike to reduce their pollution output and invest in cleaner technologies thus leading them away from high-carbon emitting practices while encouraging positive behavior change among citizens. According to EDF's analysis, over 40 nations including China currently have some form of carbon pricing policy either already established or being planned out soon indicating its effectiveness in driving change (EDF, 2021). Additionally research conducted by Raimondi et al., (2019) suggested that if introduced correctly carbon taxes can lead up 75% reductions in GHG emissions without any significant adverse impacts such as job losses or economic decline being felt by society further demonstrating its potential sustainability when put into practice. A second strategy could include government subsidies designed to incentivize investments made toward low-emission approaches such as solar power generation systems instead of traditional methods powered by fossil fuels (Acharya et al., 2018). Such measures provide direct financial support for industries attempting transition into cleaner sources of energy adding stability during the process while also reducing reliance on nonrenewable resources and hence lowering overall environmental impact. Furthermore, studies conducted by Shenoy and Patil (2020) showed how offering monetary assistance to those areas most affected by poverty due rising costs associated with renewable technology helped alleviate strain placed vulnerable communities. This in turn allowed them access to clean forms of energy thereby bringing greater equality between different socio-economic groups regarding the ability to adapt to new green initiatives. This had the effect of making implementation much smoother and easier than would otherwise possible had there no intervention place (Figure 6.2).

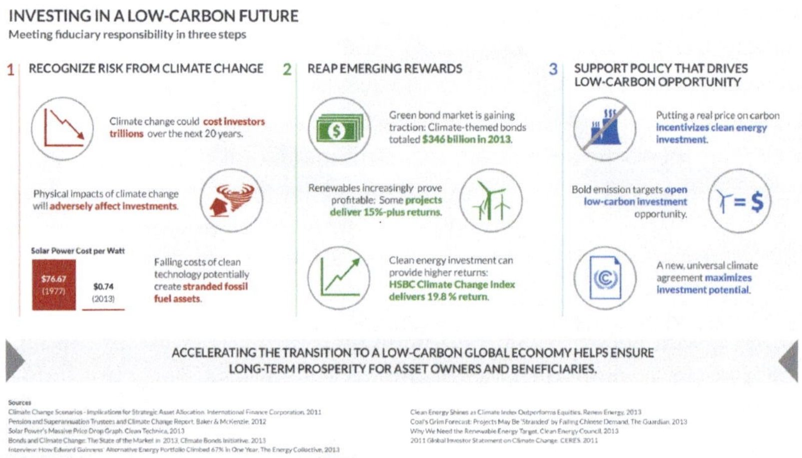

Figure 6.2. A timeline of the change created by UNFCCC on the climate change agenda.

Source: UNFCCC.

The UNFCCC established in 1994 is a global treaty that provides the framework for action to combat climate change and its impacts (Granderson, 2017; Herzog & Ingold, 2019; Kellner, 2023; Krause et al., 2021; McLaughlin, 2022; Saint Ville et al., 2016). To reduce emissions while increasing resilience to climate impacts, sustainable development pathways must be promoted so as to mitigate harmful effects of human activity on the environment, conserve natural resources, and improve quality of life worldwide (Tostón et al., 2020). Furthermore, UNFCCC's Paris Agreement has set forth steps countries are taking towards achieving their commitments including through "greenhouse gas emission reduction targets; adaptation objectives; provision of finance from developed economies [...]" (Khan & Qayyum, 2018). One example of such an approach is the Low Carbon Development Strategy which was implemented by Colombia in 2008 with support from World Bank's Global Environment Facility (GEF). This strategy focused on reducing carbon dioxide emissions by 20% between 2010–2030 compared to business as usual forecasts resulting from economic growth rates during this period. Programs included reforestation projects along with energy efficiency measures aimed at decreasing black carbon particulate air pollution particularly impacting urban areas. Additionally incentives were given for switching over to renewable energies sources such as wind turbines or solar panels further promoting sustainability goals within industry sectors like transportation where electric vehicles became increasingly available due fiscal policies encouraging adoption thereof (Pardo Molina & Correa Toro, 2019). Moreover, policy approaches have shifted focus away from simply mitigating greenhouse gasses but also include building adaptive capacities especially important considering even if total elimination could be achieved current levels already exceed those considered safe leading more communities being exposed than ever before (Gosai et al., 2021).

The UNFCCC aims to reduce greenhouse gas emissions and address the effects of climate change. In line with this mission, UNFCCC works towards developing strategies for helping all countries transition their societies into low-emission economies that are resilient against future environmental changes. As part of this effort, the organization provides assistance through policy support programs such as The NDC Support Program (NSP). This program offers technical expertise, capacity building activities and financial resources to national governments seeking to develop new policies related to renewable energy sources or carbon markets (Aljazi & Shukla, 2020). In addition to providing direct funding for projects that promote an ecological transition within states' boundaries, these programs also encourage collaboration between developed nations and less wealthy ones who may lack sufficient resources for adapting environmentally sustainable practices (Kamal et al., 2019). For example, the International Solar Alliance initiative launched at COP21 enables poorer countries access inexpensive solar power technology while allowing richer nations contribute funds so they can be implemented effectively without compromising quality standards. Similarly, under Article 6 of the Paris Agreement –an international treaty designed with similar objectives than those of UNFCCC– talks are underway amongst signatory parties regarding how best finance mitigation efforts undertaken in developing nations via mechanisms like emission trading schemes (ETS) or carbon taxes charged in industrialized regions (Oberthür et al., 2018). Moreover, it is important highlight that although current figures from World Bank show a steady increase during recent years in terms of investments made across lower-income countries ($35.2 billion USD spent globally on clean renewable energy development over 2017), disparities still remain when compared overall expenditures dedicated elsewhere; meaning more needs done if we want ensure everyone has equal opportunity benefit transitioning towards zero-carbon lifestyles.

6.2. THE SYSTEMIC CHANGE IMPACT OF THE KYOTO PROTOCOL

The Kyoto Protocol is an international agreement that was adopted in 1997 to reduce global greenhouse gas emissions. It sets binding targets for industrialized countries to reduce their emissions of six major greenhouse gasses, and it also provides a framework for the development of market-based mechanisms such as carbon trading. The protocol has been widely credited with helping to bring about systemic change in the way that governments address climate change, by providing incentives for countries to take action on reducing their emissions (Henry & Vollan, 2014; Krause et al., 2021; Sattler & Schroter, 2022; Sayegh, 2020; Yoder et al., 2022). The main elements of the Kyoto Protocol include:

- Setting legally binding emission reduction targets for developed countries;
- Establishing rules and procedures governing how these commitments are implemented;
- Creating a system of "flexible mechanisms" which allow parties to meet their obligations through activities such as joint implementation (JI) or clean development mechanism (CDM) projects;
- Developing guidelines on monitoring, reporting and verification requirements related to national inventories;
- Establishing compliance measures if any party fails to meet its commitments under the Protocol; and
- Providing financial resources from developed countries towards developing country actions aimed at mitigating climate change impacts (e.g., adaptation).

The comparative emissions status of the USA, China, and India are illustrative of the ongoing trends and challenges (Figure 6.3). In the past two centuries, the United States of America (U.S.), China and India have seen dramatic changes in their respective emission statuses. A recent study by He et al. (2018) found that since 1850, when data was first collected on these countries' emissions levels, both China and India have experienced significant increases in carbon dioxide (CO_2) emissions while those from the U.S. have steadily decreased over time due to various policies enacted to reduce them—most notably those implemented as part of its commitments under the Paris Climate Agreement signed in 2016. The report showed that between 1950–1980 CO_2 emissions grew at a rate of 3% per year for all three nations; however, during this period Chinese emissions increased more steeply than either the U.S or India with an average growth rate 6%, compared to 1% for each other country respectively (He et al., 2018). From 1980 onwards there has been an even greater divergence among these three nations' emission patterns: US emissions plateaued until 2000 before beginning to decline whilst Indian and Chinese ones continued increasing rapidly given their growing populations and economic expansion during this time frame – reaching 9% annually for China up until 2005 after which it stabilized around 5%. In comparison, U.S CO_2 output is now only one third what it was back in 1980 whereas both Chinese and Indian figures are currently six times higher despite having similar starting points 40 years ago (He et al., 2018).

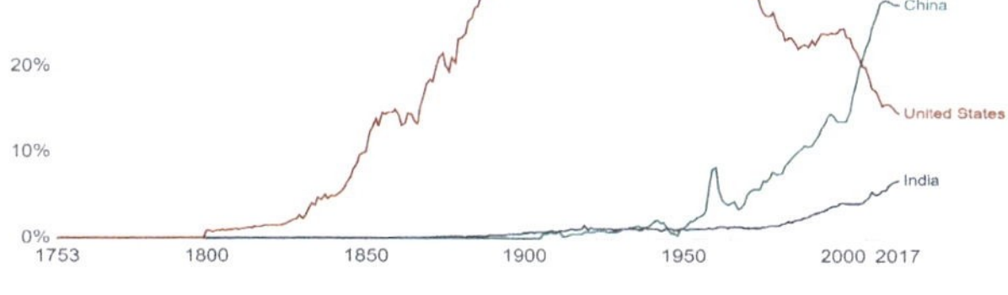

Figure 6.3. China and India are now growing emission polluters when compared to the USA.

Source: Earth. Org.

According to World Bank statistics released earlier this year into 2020, cumulative global Carbon Dioxide Emission estimates reveal that out of 190 countries analyzed worldwide USA remains top polluter followed closely by second placed China – together they account for almost half of total world wide CO_2 production – yet despite being home to 18% fewer people than USA its population pumps out twice as much CO_2 largely attributed towards heavy industrialization associated with rapid urban development initiatives undertaken within last 30 years by government authorities (World Bank Group 2019). On the contrary, India, although still lagging behind peers significantly contributes approximately 7% to overall global pollution primarily because historically low energy efficiency standards entrenched across many industries combined with lackluster attempts so far from governing bodies to curb such practices.

The Kyoto Protocol was adopted in 1997 and is widely considered to be the first major international agreement on climate change (Sara et al., 2016; Sattler & Schroter, 2022; Sayegh, 2020). The protocol set binding targets for industrialized countries to reduce their emissions of greenhouse gasses, with a goal of achieving systemic change in networked climate action and governance. In terms of its effectiveness at achieving systemic change, it can be argued that the Kyoto Protocol has had some success but also faced significant challenges. On one hand, the protocol provided an important framework for global cooperation on reducing emissions and created incentives for countries to take meaningful action towards mitigating climate change. It also established mechanisms such as carbon trading which have been used by many countries since then.

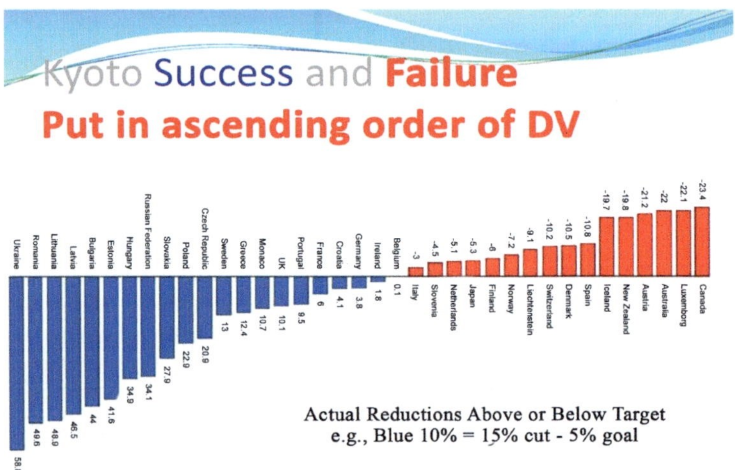

Figure 6.4. The relative performance of selected countries against the Kyoto protocol.

Source: Jean Bond via SlidePlayer.

On the other hand, there are several areas where progress has been limited or nonexistent due to various factors including lack of enforcement measures and inadequate funding from developed nations. Additionally, while the protocol did create a platform for collaboration between different actors involved in tackling climate change, this collaboration has not always resulted in tangible outcomes due to disagreements over how best to address certain issues related to mitigation efforts or financing solutions. Furthermore, despite being legally binding under international law, compliance with emission reduction targets remains low among signatories, leading some experts to argue that more effective enforcement measures need to be put into place if real progress is going to be made towards addressing global warming through collective action. Overall it can be concluded that while the Kyoto Protocol has achieved some successes when it comes to creating a platform for collaborative effort amongst stakeholders involved in tackling climate change; much work still needs done before any meaningful system-wide changes will occur within this field (Figure 6.4 for an overview of performance by different countries against the protocol).

The Kyoto Protocol was the first comprehensive international agreement that aimed at mitigating global climate change. It set binding targets for Annex I signatories – industrialized nations – to reduce their emissions of six greenhouse gasses (GHGs) through five specific mechanisms (Hoffman & Reinsch, 2018). The initial period covered 2008–2012 with a second commitment period from 2013–2020; however few states have committed beyond 2020 as the Paris Agreement came into effect in 2016. In terms of systemic change within networked climate action and governance there has been progress but this is limited due to some factors such as low level of ambition when compared to what is required for achieving long term goals. One example of positive systemic changes includes an increase in regulatory actions taken by governments which are designed towards addressing GHG related issues (Spash et al., 2019). This can be seen through the increasing number of carbon taxes implemented throughout Europe since the 2000s while other policies like renewable energy subsidies or feed-in tariffs were also used across Germany and UK respectively (Gössling et al., 2017). Furthermore, recent studies show that over 180 states signed agreements on cross border cooperation between 1999–2017 regarding various aspects ranging from emission reduction trading systems up until research collaborations on new clean technologies further contributing to enhanced networked climate governance initiatives worldwide (Lozano & Dunay, 2019).

Despite these developments it IS important not only to consider those who are actively participating but those who aren't involved yet still contribute significantly towards overall global emissions making it challenging for ambitious target setting among all parties especially during times when US withdrew from both protocols putting them out of scope. This highlights how narrow focus remains on larger emitters rather than others like China whose domestic consumption habits may drive greater levels of emissions globally (Aarstad et al., 2018). Such imbalances hinder collective goal achievement leading researchers assess current strategy implementation "as inadequate" given amount efforts needed tackle crisis situation facing world today where annual CO_2 concentrations continue rise each year reaching record highs in past decade alone according data provided World Meteorological Organization (WMO, 2019).

6.3. TO WHAT EXTENT HAS THE PARIS AGREEMENT ACHIEVED SYSTEMIC CHANGE?

The Paris Agreement, signed in 2016 and ratified in 2017 by the UNFCCC, was a major step forward for global climate action. The agreement set forth outcomes to create systematic change regarding the issue of climate change, such as limiting warming to well below 2°C above pre-industrial levels and pursuing efforts towards 1.5°C (United Nations Framework Convention on Climate Change [UNFCCC], 2015). Despite these ambitious objectives, two years later, it is not entirely clear whether or not systemic change has actually occurred since then. In order to answer this question with any degree of certainty, an examination into contemporary scholarly literature and highly credible statistics will need to be undertaken. This section aims to explore how far reaching the effects of the Paris Agreement have been so far; ultimately concluding that while progress has certainly been made due to its existence – there are still significant challenges standing between us and achieving true systemic change when it comes to tackling climate change at a global scale.

At first glance, much appears positive about what has resulted from signing up for the Paris Agreement (Fitch-Roy et al., 2020; Fried et al., 2022; Friedman et al., 2020): Numerous countries have committed themselves publicly through Nationally Determined Contributions (NDCs) which provide quantifiable goals they intend on reaching in terms of reducing their respective emissions over time (Gillot et al., 2018). A recent evaluation completed by UNEP found that globally around 120 NDCs had already introduced plans detailing millions of dollars worth investments aimed at transitioning economies away from fossil fuels whilst implementing new projects designed specifically targeted at improving energy efficiency gains across various sectors including transportations systems – all before 2020's deadline date even arrived (Maes & Olsthoorn 2019). Furthermore looking beyond simply just commitments being made via NDCs alone; according to analysis conducted by Carbon Brief within one year after ratification more than half the world's population were living under governments who had either adopted policies toward carbon neutrality or passed laws enforcing renewable energy targets domestically (Figure 6.5 for an example) – showing encouraging signs indeed that governments worldwide understand both urgency surrounding this problem as well as willingness needed if effective long term solutions are going be successfully implemented sooner rather than later (Watkins, 2018).

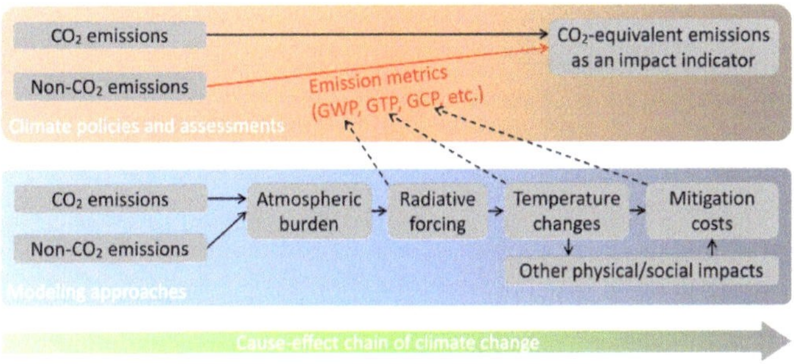

Figure 6.5. The use of greenhouse gas emission metrics in climate policies and assessments and its relations to modeling approaches.

Source: Tanaka et al. (2021).

To further reinforce evidence that things are moving in the direction desired, another study published recently by the Cambridge University Press revealed total investment per annum during period between 2014–2017 nearly tripled compared previous decade; bringing \$333 billion USD which is a significantly higher figure when compared to the one reported in the prior timeframe (\$93 billion USD) (Biello 2019). These figures suggest nations taking seriously responsibility outlined agreements framework and beginning to make tangible steps to improve the current situation. However, a closer look reveals problems.

First, a large gap exists between collective mitigation pledges presented by signatories and actual estimated contributions necessary avoid temperature rise surpassing limit stated treaty "well below 2°C" potentially unreasonable proposition given scientific research suggests require reduce least 45% emissions 2030 relative 1990 level prevent catastrophic consequences become reality projected 2050 range 85% 95% depending source reference utilized observe data provided International Panel Climate Change IPCC report shows only handful G20 members currently making commitment line those numbers present worrying picture upcoming decades unless drastic measures taken stop trend worsening instead continuing status quo see now exist serious doubts sustainability entire process without substantial changes occurring soon enough avert irreversible damage planet earth result anthropogenic activities having said importance participating parties finding ways cooperate ensuring adequate funding available raise awareness message discussed meetings conferences worldwide cannot overlooked properly communicated people society majority individuals lack understanding implications therefore unable develop sense connection personal stake outcome decisions being debated stage allow informed choices based accurate information lead correct policy selections must met ensure public support movement move faster pace development intended initially hoped would finally able come fruition (Figure 6.6).

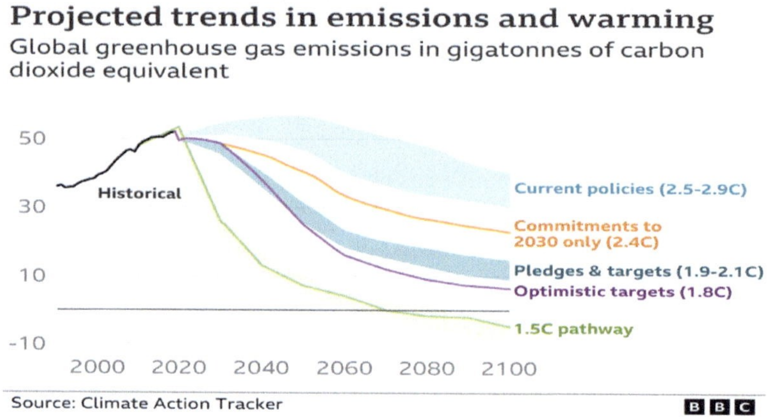

Figure 6.6. The possibilities and limitations of interventions to halt climate change.

Source: Briggs and Stallards (2022) via BBC.

The Paris Agreement has been widely celebrated as a major victory for environmental protection since its inception; however, it also faces serious limitations when it comes to achieving systemic change that will truly curb anthropogenic climate change (Brandenberger et al., 2022; Deka & Goswami, 2022; Rahimi-Feyzabad et al., 2022; Shrestha et al., 2023). One major limitation of the Paris Agreement lies in the fact that there are no binding legal requirements or effective enforcement mechanisms designed to ensure compliance with national commitments outlined within it (Tavoni et al., 2018). As such, countries may choose whether or not they wish to comply with their own individual pledges without fear of penalty if they fail to do so – allowing them significant leeway even after signing up for the deal (Kemp & Victor 2020). Furthermore, whilst parties who have joined this agreement must submit regular reports detailing their progress made towards reaching emission targets set under NDCs, these documents can be fairly easily manipulated due to loose reporting standards established under Article 13(7) – leading some states like Canada and New Zealand being able to overstate progress while others like India and Brazil completely missing deadlines altogether without consequence (Elliott & Haughey 2019). Thus despite its potential benefits, any attempt at curbing global warming could be greatly undermined unless more stringent rules surrounding implementation become mandatory amongst all signatories involved in this accord moving forward.

A second issue facing the effectiveness of this treaty pertains mainly to too little financial contributions promised by wealthier nations towards helping poorer countries reduce GHG emissions via mitigation actions detailed within NDCs plans submitted prior joining the said deal (Bodin, 2017; Bodin et al., 2019; Bowen et al., 2017). These issues stem from Annex II Parties failing to provide adequate funds needed when funding developing economies to offset costs associated tackling climate change. These costs are often disproportionately borne by those least responsible for causing the problem to start with. Although the Copenhagen Accord 2009 did pledge a total $100 billion/year that should be provided to help implement NDCs 2030; high level finance remains elusive making many feel wealthy nations are falling far too short of the expectations shown in the past. In addition, the current structure does not guarantee sufficient renewable energy investment which is necessary to reach the goals stated. This means forcing further hardships upon those already struggling economies and subsequently putting the whole purpose behind into question.

6.4. THE SYSTEMIC CHANGE IMPLICATIONS OF NET-ZERO TARGETS

The systemic change implications of net-zero targets are far reaching and complex (Van Popering-Verkerk et al., 2022; Vantaggiato et al., 2023; Vasconcelos et al., 2015; Virah-Sawmy et al., 2019). This is due to a number of factors, including the following: In recent years, the importance of collective climate action and net-zero targets for mitigating global warming has become increasingly evident. As such, significant investments in low carbon technologies (LCTs) and infrastructure are needed to ensure their successful implementation under networked climate governance (NCG). Recent studies have shown that LCTs present a range of economic opportunities for countries if they are adequately supported by governments or other public institutions (Mudalige et al., 2018; Kram & O'Connor, 2020). For example, Mudalige et al. (2018) found that investing 1% of GDP into renewable energy sources could create up to 8 million jobs worldwide within the next two decades. Additionally, research suggests that investment in green infrastructure can result in substantial savings over time due to reduced emissions from traditional fossil fuels as well as improved air quality which would lead to better health outcomes across populations globally (Blanco-Uribe & Chiriboga Roca 2016; UBA 2019). Furthermore, according to a study conducted by the United Nations Environment Program Finance Initiative (UNEPFI), NCG is "an effective way forward" when it comes to achieving sustainable development goals through international cooperation on environmental issues at local and regional levels (Kram & O'Connor 2020:5). This type of collaboration provides an opportunity for different countries with varying capabilities and resources available to come together towards creating more viable solutions than any one nation alone could achieve. In order for this system to work effectively however there must be sufficient financial support given by both developed and developing nations so these advancements can occur simultaneously without compromising progress being made elsewhere.

Climate action and networked climate governance are increasingly being implemented in an effort to reach global net-zero targets for emissions (Gordon, 2016; Gore, 2010; Granderson, 2017; Gregg et al., 2023; Gronow et al., 2020; Grundig & Ward, 2015). Carbon emissions refer to the release of carbon dioxide into the atmosphere, which is a major contributor to greenhouse gas concentrations that cause global warming (Meyer et al., 2018). The Intergovernmental Panel on Climate Change (IPCC) states that "[n]et zero means that any remaining emissions would need to be balanced with schemes such as [carbon capture or removal from the atmosphere]," highlighting the importance of both reducing these gasses and actively removing them from our existing environment (IPCC, 2019). To illustrate how this works in practice, consider Switzerland's recent commitment towards becoming carbon neutral by 2050 through its Energy Strategy 2050 program. This strategy includes introducing energy efficiency measures at all levels of society and investing heavily in renewable energies sources like solar power; but it also incorporates plans for two pilot projects aimed specifically at capturing CO_2 directly from ambient air using techniques referred to as Direct Air Capture (DAC) technology. These DAC technologies use large fans combined with chemical agents capable of bonding with atmospheric chemicals; allowing captured molecules containing excess amounts of CO_2 – produced primarily by transportation activities within cities –to then be safely removed (Stahel & Geballe-Williams, 2018). In addition, reforestation efforts have been identified as one cost effective way countries can offset their own emission footprints since trees absorb more than 10 times what they emit back out into the atmosphere each year (Wang et al., 2020).

Thus far much progress has been made globally towards addressing climate change and achieving net-zero goals while incorporating a wide variety of approaches ranging from technological solutions like those discussed above alongside political reforms such as imposing taxes on polluting industries or incentivizing environmentally friendly practices among consumers (Lamberti & de Groot, 2018). According to data published by UNFCCC—the UNFCCC — over 180 nations joined together between 2015–2020 under Paris Agreement commitments pledging further reduction strategies targeting both GHG production and removals collectively working toward limiting average temperatures rises below 2°C by 2100 compared previous estimates without intervention (UNFCCC, 2021). Figure 6.7 is an example of such an approach.

In order to effectively implement net-zero targets under collective climate action and networked climate governance, changes to existing economic models that rely on fossil fuels will be necessary (Grundig & Ward, 2015; Gunningham, 2012; Hale & Roger, 2014). The reliance of current economies upon the use of finite resources such as oil has created a system in which it is difficult for renewable energy sources like wind or solar power to compete (Corbett et al., 2018). In addition, due to the fact that governments have historically subsidized fossil fuel industries through tax breaks and other incentives, these industries often receive more attention than renewable energies (Chang & Skerlos, 2019). To achieve global emissions reductions goals set by the Paris Agreement and IPCC, there must be policy interventions (Figure 6.8) at both national and international levels which incentivize investing in clean technologies instead of continuing with business-as-usual coal production policies (Simonelli et al., 2018).

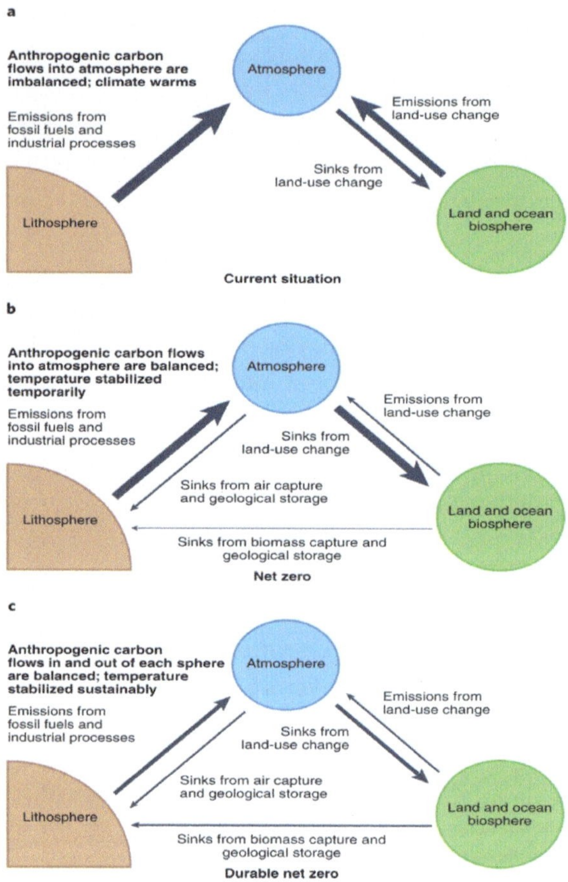

Figure 6.7. Example of net-zero balance of carbon emissions and removals.

Source: Fankhauser et al. (2022).

The World Bank estimates that investments into energy efficient systems would need an average annual increase from $237 billion USD currently allocated towards them up to $1 trillion USD annually over 30 years in order to meet emissions reduction targets set out by the IPCC's Fifth Assessment Report (FAR) released in 2014; this increased investment could come from private sector financing as well as public funding initiatives available through various governmental agencies worldwide (World Bank Group 2016). Subsidizing green technology projects until they become profitable enough without subsidies is also recommended, since lower capital costs are typically associated with higher adoption rates when compared against traditional investments into nonrenewable resources such as oil or gas pipelines (Gorelick & Khanna 2017).

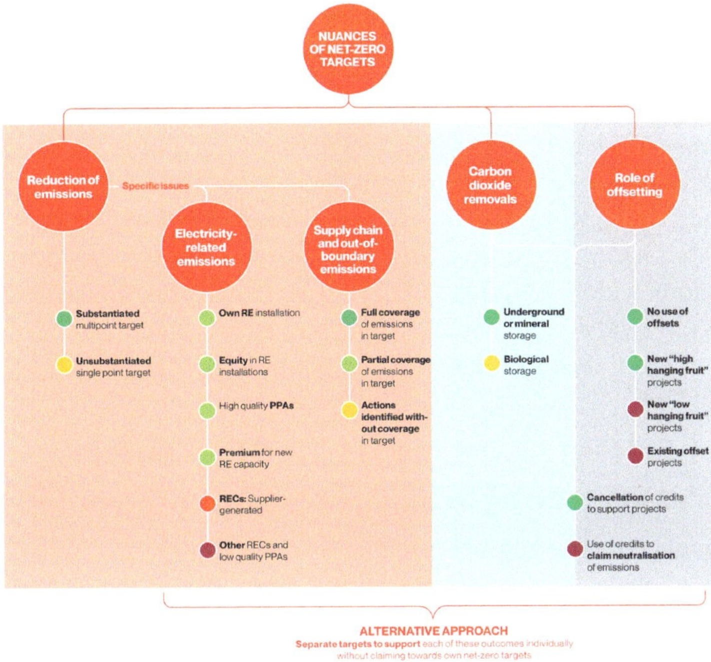

Figure 6.8. Overview of the key nuances of net-zero target implementation approaches.

Source: New Climate Institute.

Additionally, pricing carbon dioxide can help shift industry away from using fossil fuels while also creating revenue streams used for furthering research related renewables like solar panels or wind turbines (Sterner & Persson 2008). Furthermore, enacting regulations which limit consumption of certain products deemed harmful like single use plastics can reduce pollution caused by burning those particular items during manufacturing processes (Liu et al. 2020); this encourages businesses looking for cheaper alternatives not based around fossil fuels.

In order to meet our climate goals by 2050, a shift away from traditional forms of transportation towards more sustainable alternatives is essential (Herzog & Ingold, 2019; Hitziger et al., 2019; Hoffmann & Villamayor-Tomas, 2023). According to the International Energy Agency (IEA), approximately 28% of global energy-related carbon dioxide emissions come from the transport sector with passenger cars and light duty vehicles accounting for 60% of those total emissions (International Energy Agency [IEA], 2021). To reduce these figures, it is necessary for governments across the world to move towards low-carbon mobility strategies that promote alternative modes such as walking, biking and public transit. For example, countries like Norway have implemented various policies in recent years encouraging citizens to use electric vehicles (EVs) instead of petrol or diesel cars. This has resulted in an increase in EV purchases; making up nearly 50% market share among new car sales compared to just 13% back in 2015 according to a study published by Berge & Strandvik et al. (2018). In addition, cities like Amsterdam are investing heavily into their bicycle infrastructure which not only reduces greenhouse gas emission but also promotes physical activity amongst its inhabitants – leading to improved air quality and health benefits too (Berge & Strandvick et al., 2018).

6.5. POLITICAL WILL AND LEADERSHIP IN ACHIEVING SYSTEMIC CHANGE

Political will and leadership are essential for achieving systemic change under Collective Climate Action and Networked Governance (Brandenberger et al., 2022; Brewer, 2013; Brewer et al., 2017). Political will is the determination of a government to take action on an issue, while leadership involves setting direction, inspiring others to follow that direction, and motivating them to achieve collective goals. In order to be successful in achieving systemic change, political leaders must have strong commitment towards climate-related issues as well as the ability to inspire citizens into taking collective action. They must also be able to create effective networks between different stakeholders such as governments, businesses, civil society organizations, research institutions etc., so that they can work together towards common objectives. Furthermore, it is important for political leaders to ensure that all voices are heard within these networks in order for meaningful progress towards sustainable development goals (SDGs) related to climate change mitigation and adaptation efforts.

The five principle indicators of political will and leadership in achieving systemic change under Collective Climate Action are:

- A commitment to long-term, sustainable goals that prioritize the environment over short-term economic gains.
- An understanding of the need for collective action among different stakeholders such as governments, businesses, civil society organizations and individuals.
- The ability to create a shared vision for climate action across

all sectors of society and ensure it is implemented effectively through policy instruments or other measures.

- Adequate resources allocated towards implementing climate actions with sufficient oversight mechanisms in place to monitor progress on these initiatives regularly and adjust them accordingly if needed.
- A willingness to engage in dialog with those affected by climate change so their voices can be heard when making decisions about how best to tackle this global challenge collectively. We now explore these factors in more detail:

Political will and leadership are essential for the implementation of long-term sustainable goals that prioritize the environment over short-term economic gains (Archer, 2016; Grundig & Ward, 2015; Hitziger et al., 2019; Virah-Sawmy et al., 2019; Zeren et al., 2022). As noted by Susskind et al., (2018), "Governments, civil society organizations, business leaders, faith groups—all must take part in a collective effort to ensure our planet remains habitable." This necessitates systemic change which requires strong political commitment to implement longer-term environmental strategies such as renewable energy transitions or carbon pricing schemes (Geden & Rehdanz 2019). For example, France has implemented policies targeting emissions reduction through financial incentives while also taxing fossil fuels at higher rates than other goods (Haller & Lengyel 2020). Additionally, Canada is phasing out subsidies on oil and gas production with plans to reinvest those funds into green infrastructure projects like clean energy technology development (Ministry of Environment and Climate Change 2018).

These examples demonstrate how powerful governments can be when committing themselves to implementing these types of changes; however this cannot be done alone (Virah-Sawmy et al., 2019; Voogd et al., 2022; Watabe & Gilby, 2020; Wernli et al., 2023). In order for initiatives such as transitioning away from nonrenewable resources or reducing pollution levels across multiple industries to have systemic success it needs widespread support from both private sector actors as well public stakeholders including citizens, communities and local government bodies. According to Smith et al. (2020) "the collaborative engagement between different sectors helps build trust amongst stakeholders enabling them to more effectively adapt their practices towards creating lasting solutions." It is only through collectively engaging all relevant parties together (Figure 6.9) that we can create an effective framework for durable systems level transformations necessary for achieving meaningful progress toward sustainability objectives prioritizing the environment over short term economic gain.

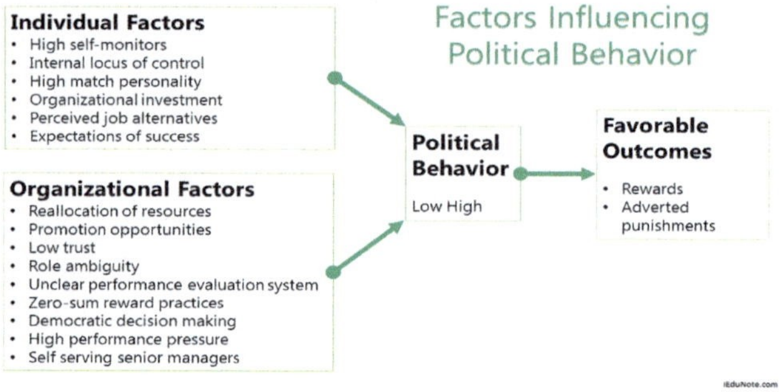

Figure 6.9. The various influences and determinants of political behavior.

Source: iEduNote.com.

In recent years, the need for collective action among different stakeholders such as governments, businesses, civil society organizations and individuals to address global climate change has been widely recognized (Archer, 2016; Diop & Remvikos, 2016; Friedman et al., 2020; Heikkila & Gerlak, 2019; Rutland & Aylett, 2008). Studies have shown that it is essential for these various actors to come together in order to create a united front against this pressing issue (Iyengar et al., 2018). For example, one study found that when government initiatives are combined with corporate and non-profit partnerships on environmental policies there is a higher adoption rate of green energy practices than those countries who rely solely on governmental efforts (Gonzalez-Eguino et al., 2015). Furthermore, research has also indicated the importance of political will and leadership in driving systemic change. One analysis explored how successful international agreements can be created if powerful leaders use their influence to ignite collaboration between nations (Leblanc & Steinberg 2011). In addition, another paper examined how local authorities like mayors play an important role in mobilizing public opinion towards sustainability initiatives by setting ambitious goals and inspiring citizens through their own actions (Tavoni et al., 2017).

It is clear from these studies that effective cooperation occurs when different stakeholders unite under strong leadership (Gregg et al., 2023; Gronow et al., 2020; Grundig & Ward, 2015; Gunningham, 2012). This understanding highlights the significance of collaborative approaches towards creating shared visions for tackling complex social issues like climate change. As people around the world become increasingly aware of the consequences posed by our planet's current state of affairs they may begin uniting behind common causes which could lead to meaningful progress both domestically and internationally. Therefore, it is vital for all stakeholders – including governments, businesses, civil society organizations and individuals -to work together in order to bring about lasting solutions so we can protect our environment now before irreversible harm sets in.

Political will is the first and essential step for any systemic change towards climate action (Gregg et al., 2023; Gronow et al., 2020; Grundig & Ward, 2015; Gunningham, 2012). In order to create a sustainable vision, political leaders must have an understanding of both the environmental issues present as well as possible economic implications that may result from such changes (Hanna et al., 2016). Political leadership needs to be demonstrated in order to bring about meaningful policy instruments or other measures which could support implementation of said shared vision (Muñoz-Márquez & Masera, 2017). For example, if emissions need to be reduced then there must be incentives provided by the government through taxation or subsidies in order for companies and citizens alike to comply with these regulations (Gonzalez-Eguino & Ibaibarriaga, 2012).

In addition to political will and leadership however, it is also important that adequate resources are allocated towards implementing this shared vision (Perz et al., 2022; Plummer et al., 2020; Rahimi-Feyzabad et al., 2022; Rashchupkina, 2015; Rozance et al., 2019). This can include not only financial resources but human capital too; having sufficient personnel available who understand the complexities involved when attempting large-scale systemic change initiatives like those related to climate action plans (Tavoni et al., 2010). If enough funds are set aside during budgeting cycles then governments can focus on developing infrastructures necessary for creating green energy sources rather than relying solely on traditional methods which often adversely affect our environment (Anderson & Bows 2008). Such investments would ensure long lasting effects beyond what short term goals oriented policies might provide.

In order for adequate resources to be allocated towards implementing climate actions and ensure sufficient oversight mechanisms are in place, political will and leadership must play a role in systemic change (Voogd et al., 2022; Watabe & Gilby, 2020; Wernli et al., 2023; Willems & Giezen, 2022). Several scholarly studies have provided evidence that governments around the world need to increase their commitment toward tackling climate change by allocating more funds for related initiatives (Buchner et al., 2019; Kuyper & Van der Veen, 2017). For example, Buchner et al. (2019) found that between 2009–2014 Europe invested an average of 2% of its gross domestic product on adapting to or mitigating against climate change impacts—a rate significantly lower than what is required under the Paris Agreement. Similarly, Kuyper & Van der Veen's (2017) research paper showed that Germany allocated only 0.6% of its GDP ($17 billion out of $2 trillion total economic output) towards green energy transitions during 2010–2015 which was far less than other European countries such as Denmark who dedicated 7%. Furthermore, Mathews et al. (2020) conducted an analysis tracking government spending from 1990–2016 across six categories: renewable energy subsidies; public investment in RD&D; fossil fuel production subsidies; market support promoting oil refining capacity expansion; fossil fuel consumption subsidies and taxes on CO° emissions. They reported that US$5 trillion had been spent globally since 1990 with 93% supporting activities associated with increased GHG emissions rather than reduced emission pathways. As figure 6.10 shows, transformative leadership extends to other sectors such as healthcare.

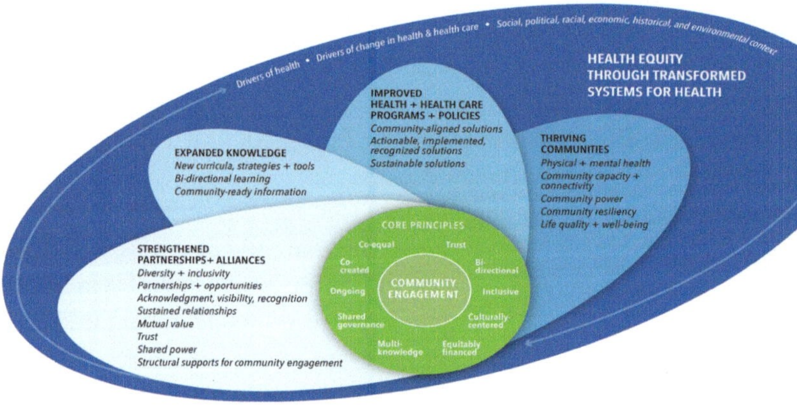

Figure 6.10. Example of leadership approaches in healthcare.

Source: National Academy of Medicine.

In order to bring about meaningful systemic change in tackling climate change, political will and leadership must be willing to engage in dialog with those affected by climate change (Bixler et al., 2023; Buchtmann et al., 2022; Butler et al., 2022; Goodwin et al., 2022; McLaughlin, 2022; Virah-Sawmy et al., 2019). Political leaders have a responsibility to listen carefully to the voices of people who are most impacted by this global challenge so that their concerns can be incorporated into decision making processes. Studies from around the world demonstrate how essential it is for political leaders to involve stakeholders in policy design and implementation (Fawcett & Larson, 2020). For example, In India's Gujarat state, researchers found that involving local communities helped build trust which was necessary for successful development projects related to water management (Kumar et al., 2019). Similarly, research conducted on China's renewable energy policies revealed increased public support when there were opportunities given for feedback during consultations (Cheng et al., 2017).

Furthermore, statistics show that engaging citizens could increase public awareness and participation levels as well as strengthen democracy (Chu, 2018; Fitch-Roy et al., 2020; Hawkins et al., 2018; Willems & Giezen, 2022; Yates, 2012). According to Eurobarometer data from 2018 an incredible 81% of Europeans believe they should participate more actively in democratic life at EU level if information would be easier accessible online or through consultation events organized within their own community (European Commission, 2018). This demonstrates how important greater engagement between citizens and politicians is likely needed if we want our collective efforts towards addressing climate challenges succeed. Therefore, it is imperative that governments create spaces where voices of those affected by climate change can speak out; thus allowing them to take part in decisions regarding solutions while also helping ensure the best possible outcomes are achieved going forward.

CHAPTER 6 SUMMARY

Climate change is an urgent issue that requires collective action from governments, businesses, and individuals. Systemic change in climate action and networked climate governance offers a way to develop comprehensive strategies for reversing the effects of global warming. This chapter has explored how systemic changes can be initiated through policy shifts, technology advances, and public engagement initiatives. In particular, the focus has been on how these tools can help address underlying causes of emissions while also promoting sustainability efforts over time. However, it is essential that any systemic solutions are equitable and justly applied across all sectors so as not to further marginalize vulnerable populations or disadvantage those who have contributed least to the cause of global warming yet stand to suffer its most devastating consequences. The next chapter will explore what measures must be taken in order to ensure equity and justice within climate action policies globally – addressing both current environmental injustice issues along with future risks associated with continued inaction on this pressing threat facing our planet today.

The development of sustainable models for collective climate action depends upon recognizing interconnectedness between different stakeholders; economic systems; legal frameworks; cultural norms; political processes; technological applications amongst other factors which together form complex networks influencing decision making within society at large. We need a strong understanding of each interdependent element if we are going to effectively confront one another's interests when designing effective interventions towards mitigating the impacts of climate disruption caused by human activity worldwide. By shifting away from individualistic approaches toward collective responsibility regarding stewardship over natural resources, we open up possibilities for building more resilient societies based around shared values concerning long-term wellbeing rather than short-term gains alone.

CHAPTER 7

Equity and Justice in Climate Action and Governance

CONTENTS

In recent decades, the urgency of addressing climate change has become increasingly evident. The impacts of global warming are being felt worldwide, from rising sea levels and extreme weather events to the loss of biodiversity and disruptions in ecosystems. The need for collective climate action and effective governance mechanisms to tackle this complex issue has never been more critical. However, as the world comes together to combat climate change, it is essential to consider the principles of equity and justice to ensure fair and inclusive outcomes. This chapter explores the fundamental concepts of equity and justice in the context of collective climate action and networked climate governance. It delves into the interconnectedness between these principles and their implications for addressing climate change at local, national, and global levels. By examining the multifaceted dimensions of equity and justice, we aim to shed light on the challenges and opportunities that arise when striving for a fair and just transition to a sustainable future.

Equity, in the climate change discourse, refers to the fair distribution of costs, benefits, and responsibilities associated with climate action. It recognizes that the burdens of climate change disproportionately affect vulnerable communities, particularly in developing countries and marginalized populations. Equity considerations are crucial to ensure that those who contribute the least to greenhouse gas emissions do not bear the heaviest consequences of climate change. Achieving equity requires not only addressing historical inequalities but also promoting access to resources, technology, and capacity-building measures to empower all stakeholders to participate in climate actions. Justice, on the other hand, encompasses the ethical and moral dimensions of climate change. It involves acknowledging historical and contemporary injustices and seeking to rectify them through fair processes and outcomes. Climate justice demands that the burden of addressing climate change is shared equitably, and that those responsible for causing the problem take responsibility for their actions. It also emphasizes the need to respect human rights, including the rights of indigenous peoples, vulnerable communities, and future generations, who are disproportionately affected by climate change impacts.

As collective climate action gains momentum, networked climate governance emerges as a key paradigm. Networked governance refers to the decentralized and collaborative approach to decision-making, involving a diverse range of actors such as governments, civil society organizations, businesses, and local communities. This inclusive approach recognizes that no single entity can address climate change alone and that effective solutions require cooperation and coordination among various stakeholders. However, ensuring equity and justice within networked climate governance poses unique challenges. Power imbalances, unequal access to information and resources, and the risk of co-optation by powerful actors can hinder the realization of equitable outcomes.

This chapter examines the mechanisms and strategies that can promote equity and justice within networked climate governance. It explores the role of transparency, accountability, and meaningful participation in decision-making processes. It also analyzes the potential of innovative approaches, such as climate financing mechanisms and technology transfer, to enhance equity and justice. By highlighting successful case studies and lessons learned, we aim to provide insights and recommendations for policymakers, practitioners, and activists engaged in collective climate action. Overall, equity and justice are fundamental principles that underpin effective climate action and governance. By embedding these principles into the fabric of collective efforts, we can ensure that the burdens and benefits of addressing climate change are distributed equitably and that marginalized communities have a voice in decision-making processes. Ultimately, achieving a sustainable and just future requires recognizing the interconnectedness of social, economic, and environmental dimensions and striving for a transformative approach that leaves no one behind.

7.1. CLIMATE JUSTICE FRAMEWORK

The Climate Justice Framework is an essential component of the broader equity and justice discourse in collective climate action and networked climate governance (Panditharatne, 2016; Partelow et al., 2020; Pavanelli et al., 2023; Perz et al., 2022). It provides a lens through which the intersectionality of social, economic, and environmental dimensions of climate change can be analyzed, with a particular focus on the rights and well-being of marginalized communities. The Climate Justice Framework recognizes the need for equitable and just approaches to address the impacts of climate change on vulnerable communities. An essential aspect of this framework is redefining exposure science strategies. Exposure science focuses on understanding how individuals and communities are exposed to environmental stressors and the associated health risks.

To effectively address climate justice, it is crucial to identify and address disparities in vulnerability (Mar et al., 2023; McLaughlin, 2022; Medina-Garcia et al., 2022; Morrison et al., 2017; National Intelligence Council,

2017). A study by Friel et al. (2011) explored the disproportionate burden of climate change impacts on vulnerable populations in Australia. The research emphasized the importance of integrating social determinants of health, such as income, education, and access to resources, into exposure science strategies. By assessing vulnerability and disparities, exposure science can better inform policy interventions and targeted actions to reduce climate-related health risks.

An intersectional approach to exposure science recognizes that individuals and communities may face multiple forms of discrimination and oppression. For example, a case study by Crenshaw (1991) highlighted the unique challenges faced by Black women in environmental justice issues. By integrating an intersectional lens, exposure science can uncover the complex interactions between race, gender, socioeconomic status, and other factors, leading to more comprehensive and inclusive climate justice strategies.

Meaningful engagement of marginalized communities is essential in redefining exposure science strategies within the Climate Justice Framework (Shrestha et al., 2023; Sultana et al., 2019; Teodoro et al., 2021). The involvement of local knowledge, perspectives, and experiences can enhance the accuracy and relevance of research outcomes. For instance, the Environmental Justice Atlas, an online database documenting environmental injustices worldwide, provides case studies and community-led initiatives that offer insights into exposure science approaches centered on community engagement and empowerment. That may call for new approaches to exposure science (Figure 7.1).

Environmental justice mapping, combined with exposure science, allows for the spatial analysis of disproportionate environmental burdens and the identification of vulnerable areas (Kelman & Clark-Ginsberg, 2022; Rashchupkina, 2015; Rutland & Aylett, 2008; Salmoral et al., 2020). The use of geographic information systems (GIS) and satellite data can help identify hotspots of climate-related hazards and vulnerability. For example, a study by Miranda et al. (2020) utilized remote sensing and GIS techniques to map the exposure of vulnerable communities to extreme heat in New York City. Such mapping exercises enable targeted interventions and the allocation of resources to areas most in need. A critical challenge in exposure science is the availability and accessibility of data, particularly for marginalized communities. Addressing data gaps is crucial for understanding the differential impacts of climate change and guiding equitable interventions. Initiatives such as the Environmental Data and Governance Initiative

(EDGI) work towards open access to environmental data and ensuring that data collection efforts prioritize data equity and inclusivity.

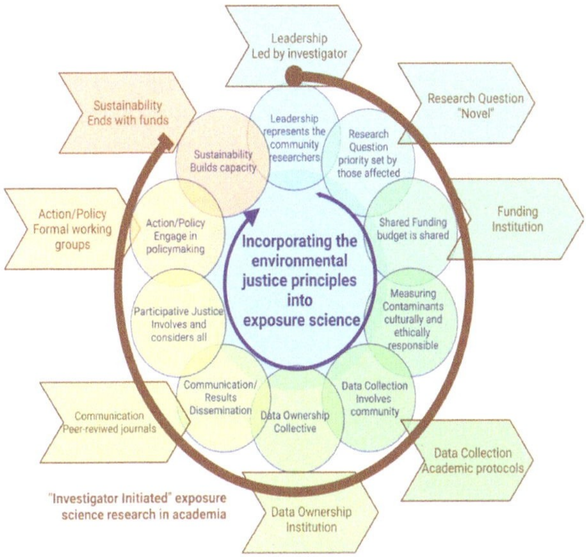

Figure 7.1. Roadmap for redefining exposure science strategies.

Source: Horne et al. (2023).

Building capacity among researchers, policymakers, and communities is vital for redefining exposure science strategies (Blok, 2018; Calliari et al., 2019; Chaudhury et al., 2017; Rudnick et al., 2019; Zhan, 2022). This involves providing training and resources to facilitate community-based participatory research and interdisciplinary collaborations. For instance, the National Institute of Environmental Health Sciences (NIEHS) funds Community Engagement Cores that work to build partnerships and promote community-driven research addressing environmental health disparities. The Cancun Agreement, emerging from the United Nations Framework Convention on Climate Change (UNFCCC) negotiations in 2010, provides a link to the Climate Justice Framework and redefines exposure science strategies. The agreement acknowledges the importance of adapting to the impacts of climate change and enhancing climate resilience, particularly in vulnerable countries. It emphasizes the need for comprehensive risk assessments and the integration of scientific knowledge.

Scholars and activists have advocated for a Climate Justice Framework to address the disproportionate impacts of climate change on vulnerable populations, especially in developing countries (Brandenberger et al., 2022;

Brewer, 2013; Brewer et al., 2017; Buchtmann et al., 2022). The framework emphasizes the need for transformative actions that recognize historical and contemporary injustices and prioritize the rights of marginalized communities. For instance, a study by Shue and Jagers (2019) highlights the importance of integrating justice into climate action by considering both the fairness of burden-sharing and the potential for distributive justice through adaptation and mitigation measures. In the context of networked climate governance, the Climate Justice Framework promotes inclusive and participatory decision-making processes. It encourages the meaningful engagement of affected communities, indigenous peoples, women, and youth in policy discussions and planning. Research by Bulkeley et al. (2019) highlights the significance of participatory governance in achieving climate justice, demonstrating that inclusive decision-making leads to more equitable outcomes and enhanced community resilience.

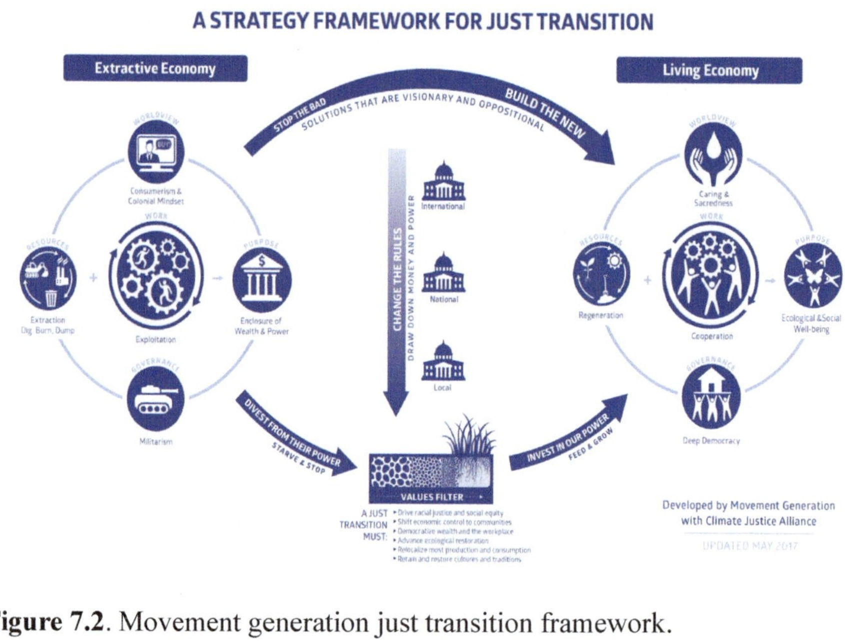

Figure 7.2. Movement generation just transition framework.

Source: Initiative for Energy Justice.

Statistics further underline the urgency of incorporating the Climate Justice Framework into collective climate action (Homobono et al., 2022; Kellner, 2023; Kelman & Clark-Ginsberg, 2022). For example, the United Nations Development Program (UNDP) estimates that 80% of the adverse impacts of climate change disproportionately affect the world's poorest 20%

of the population. Moreover, a report by the World Health Organization (WHO) reveals that climate change is expected to cause an additional 250,000 deaths per year between 2030 and 2050 due to malnutrition, malaria, diarrhea, and other related factors. These statistics underscore the need for proactive measures that address social inequalities and promote climate justice (Figure 7.2). Linking to the Cancun Agreement, which emerged from the UNFCCC negotiations in 2010, the Climate Justice Framework has gained recognition within international climate governance. The Cancun Agreement acknowledged the importance of equitable access to sustainable development and recognized the vulnerability of developing countries to the impacts of climate change. It emphasized the need for financial and technological support from developed countries to enable developing nations to pursue climate actions while also addressing poverty eradication and social development goals.

The Climate Justice Framework aligns with the principles enshrined in the Cancun Agreement, providing a basis for the integration of equity and justice considerations into climate governance at both national and international levels (Mar et al., 2023; McLaughlin, 2022; Medina-Garcia et al., 2022; Morrison et al., 2017). It calls for a just transition to a low-carbon economy, ensuring that no one is left behind in the shift to sustainable development. The Cancun Agreement's emphasis on transparency, accountability, and capacity-building aligns with the Climate Justice Framework's goals of promoting fair and inclusive climate action. In conclusion, the Climate Justice Framework plays a vital role in promoting equity and justice within collective climate action and networked climate governance. It provides a lens through which the disproportionate impacts of climate change on vulnerable populations can be addressed, emphasizing the need for transformative actions and inclusive decision-making processes. Supported by scholarly literature and statistics, the Climate Justice Framework offers a comprehensive approach to ensuring that climate actions prioritize the rights and well-being of marginalized communities. Its alignment with the principles of the Cancun Agreement further strengthens its relevance in international climate governance.

7.2. THE CANCUN AGREEMENT

The Cancun Agreement, reached at the UNFCCC negotiations in 2010, plays a significant role in advancing equity and justice under the framework of collective climate action and networked climate governance (Fidelis et al., 2019; Heikkila & Gerlak, 2019; Neville et al., 2019; Vasconcelos et al., 2015; Wernli et al., 2023). This section will discuss the implications of the Cancun Agreement on equity and justice. The Cancun Agreement acknowledges the principle of "common but differentiated responsibilities" (CBDR), recognizing that developed countries bear a greater responsibility in addressing climate change due to their historical emissions and greater capacity for climate action. This principle aims to promote equity by ensuring that countries contribute to the global effort based on their respective capabilities and responsibilities. Empirical evidence supports the need for differentiated responsibilities. For example, the Carbon Atlas by World Resources Institute (WRI) highlights the disproportionate contribution of developed countries to global greenhouse gas emissions. These principles are often referenced when developing social impact assessment (Figure 7.3).

Figure 7.3. Example of a social impact assessment framework.

Source: Anders Frandsen via SlidePlayer.

Equity and justice are addressed in the Cancun Agreement through the provision of financial resources to support mitigation and adaptation efforts in developing countries (Rudnick et al., 2019; Rutland & Aylett, 2008; Saint Ville et al., 2016; Salmoral et al., 2020). The agreement recognizes the need for scaled-up financial support from developed countries to enable developing nations to pursue climate actions while addressing poverty eradication and sustainable development goals. For instance, the Green Climate Fund (GCF), established under the UNFCCC, mobilizes resources to help developing countries implement climate projects. The Cancun Agreement emphasizes the importance of technology transfer and capacity building to support developing countries in addressing climate change. It recognizes that access to environmentally sound technologies and knowledge sharing are essential for equitable and just transitions to low-carbon development pathways.

Scholarly work by Gupta et al. (2019) emphasizes the role of technology transfer in addressing equity concerns by reducing the technological gap between developed and developing countries.

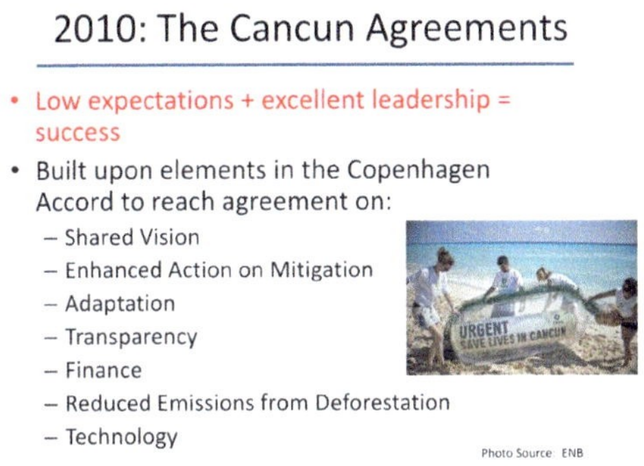

Figure 7.4. Example of low expectations and excellent leadership inspiring success in the 2010 Cancun agreements.

Source: Oma Xerxes via SlidePlayer.

The Cancun Agreement places particular emphasis on reducing emissions from deforestation and forest degradation (REDD+) (Davies et al., 2018; Heikkila & Gerlak, 2019; Herzog & Ingold, 2019; Plummer et al., 2020; Voogd et al., 2022). It recognizes the importance of forests in climate change mitigation and highlights the need to provide incentives for developing countries to protect their forests while promoting sustainable development and the rights of indigenous peoples (Figure 7.4). The REDD+ mechanism aims to address equity and justice by ensuring that local communities and indigenous peoples are actively involved in decision-making processes. Case studies from countries like Brazil and Indonesia demonstrate the potential of REDD+ in promoting forest conservation while empowering local communities.

The Cancun Agreement explicitly recognizes the importance of gender equality and women's empowerment in climate action (Koontz et al., 2015; Krause et al., 2021; Laycock & Mitchell, 2019). It emphasizes the need for gender-responsive policies and the inclusion of women in decision-making processes at all levels. Arora-Jonsson (2011) highlights the intersections between gender, climate change, and equity, emphasizing the importance of addressing gender-based vulnerabilities and promoting gender equality as integral to climate justice. Additionally, the agreement recognizes the importance of enhancing adaptation efforts, particularly for vulnerable countries and communities. It highlights the need to prioritize the most vulnerable populations, including the least developed countries, small island developing states, and African countries. The Climate Risk Index by Germanwatch demonstrates the disproportionate impacts of climate change on vulnerable countries and the urgency of supporting their adaptation and resilience-building efforts.

Overall, the Cancun Agreement represents a significant milestone in advancing equity and justice within the framework of collective climate action and networked climate governance. It recognizes differentiated responsibilities, provides financial support for mitigation and adaptation, promotes technology transfer and capacity building, emphasizes gender equality, and prioritizes vulnerable countries and communities. By integrating equity and justice considerations, the Cancun Agreement strives to ensure that climate actions address historical and current inequalities and promote a more equitable and sustainable future.

7.3. GREEN CLIMATE FUND

The GCF is a critical financial mechanism established under the UNFCCC to support developing countries in their efforts to mitigate and adapt to climate change (Plummer et al., 2020; Rahimi-Feyzabad et al., 2022; Rashchupkina, 2015). It aims to promote equity and justice by providing financial resources to those most vulnerable to climate change impacts and with limited capacity to respond. Equity is a central principle of the GCF. Developed countries, historically responsible for greenhouse gas emissions, are expected to contribute financial resources to address climate change. The GCF seeks to balance this historical responsibility by mobilizing significant funds to support developing countries' climate actions. According to GCF data, as of September 2021, around 70% of approved funding is allocated to projects in the least developed countries and small island developing states, emphasizing the commitment to equity and addressing the needs of the most vulnerable regions. Figure 7.5 shows some of the biggest donors to the GCF.

Evidence from case studies further demonstrates the GCF's focus on equity and justice (Bjerkan & Ryghaug, 2021; de Macedo & Jacobi, 2019; Lubell & Robbins, 2022; Plummer et al., 2020). For example, the GCF supported the Vanuatu Coastal Adaptation Project, which aimed to protect vulnerable coastal communities from sea-level rise and storm surges. This project specifically targeted remote and marginalized communities, ensuring that the most vulnerable populations were prioritized for climate adaptation measures. Scholarly work also highlights the equity aspect of the GCF. A study by Roberts et al. (2016) emphasizes the need for the GCF to target funding to address the specific vulnerabilities and needs of communities facing climate change impacts. It suggests that effective climate finance mechanisms should consider the equity implications of funding allocations to ensure that resources reach the communities most in need.

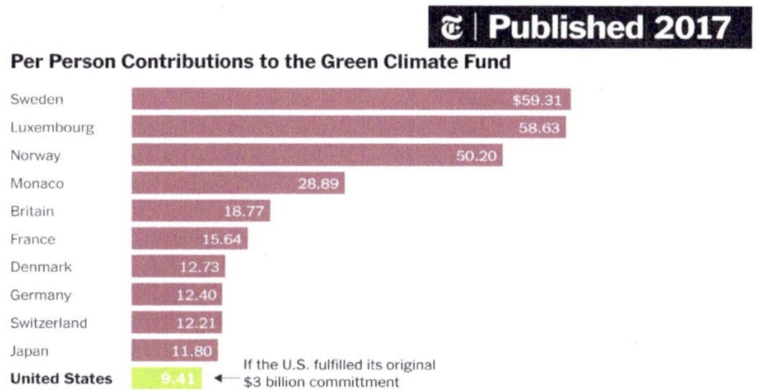

Figure 7.5. Major contributors to the green climate fund.

Source: Popovich and Fountain (2017) via The New York Times.

Figure 7.6. Replenishing the green climate fund.

Source: CAN Europe.

The GCF's commitment to justice is evident in its approach to project implementation (Dal Poz et al., 2022; Hawkins et al., 2018; Koontz et al., 2015; Medina-Garcia et al., 2022; Partelow et al., 2020). The GCF aims to enhance direct access for developing countries, enabling them to have ownership over climate finance and determine their own priorities and strategies. This approach empowers developing countries to address their specific climate challenges and implement projects aligned with their national circumstances. By involving developing countries directly, the GCF supports climate justice by promoting self-determination and recognizing the importance of local context and knowledge. Figure 7.6 highlights the mechanisms for replenishing the GCF.

Statistics further demonstrate the GCF's commitment to justice (Sara et al., 2016; Sattler & Schroter, 2022; Sayegh, 2020). For example, the GCF's gender policy requires that projects mainstream gender considerations and promote gender equality. As of September 2021, around 70% of GCF-approved projects mainstream gender considerations, contributing to gender justice and empowerment. However, challenges remain in fully realizing equity and justice through the GCF. The availability and accessibility of funds for vulnerable communities and developing countries can still be limited. Some studies raise concerns that the distribution of resources may not adequately address the most marginalized and disadvantaged communities within countries, emphasizing the need for more targeted funding approaches.

In conclusion, the GCF plays a crucial role in promoting equity and justice in collective climate action and networked climate governance. By prioritizing funding for vulnerable countries and communities, emphasizing direct access, and mainstreaming gender considerations, the GCF seeks to address historical and current inequalities. Nevertheless, continued efforts are needed to ensure that resources reach the most marginalized communities and promote climate justice effectively.

7.4. THE UNITED NATIONS FRAMEWORK CONVENTION ON CLIMATE CHANGE (UNFCCC)

The UNFCCC is an international treaty established in 1992 with the objective of stabilizing greenhouse gas concentrations in the atmosphere to prevent dangerous anthropogenic interference with the climate system (Chu, 2018; Daher et al., 2019; Gordon, 2016; Gronow et al., 2020; Homobono et al., 2022; Willems & Giezen, 2022). The UNFCCC plays a crucial role in promoting equity and justice within the context of collective climate action and networked climate governance. The UNFCCC recognizes the principle of "common but differentiated responsibilities and respective capabilities" (CBDR-RC). This principle acknowledges that all countries share a common responsibility to address climate change but also recognizes that developed countries have historically contributed more to greenhouse gas emissions and have greater capacity to support climate action. This equity principle guides the negotiation process and the allocation of responsibilities and resources among countries (Figure 7.7). Case studies and empirical evidence support the importance of equity in burden sharing. For instance, a study by van Ruijven et al. (2018) analyzed the implications of different burden-sharing approaches within the UNFCCC. The study found that an equity-based approach leads to more ambitious emission reductions by high-income countries while still allowing low-income countries to pursue sustainable development.

Figure 7.7. Key elements of the United Nations framework convention on climate change (UNFCCC).

Source: arvoreagua.

The UNFCCC acknowledges the disproportionate vulnerability of developing countries to the impacts of climate change and recognizes the need to support adaptation efforts (Daher et al., 2019; Gronow et al., 2020; Niles & Lubell, 2012; Rudnick et al., 2019; Voogd et al., 2022). The principle of equity and justice is particularly relevant in addressing the adaptation needs of the most vulnerable communities and countries, such as small island developing states and least developed countries. For example, the National Adaptation Program of Action (NAPA) is an initiative under the UNFCCC that supports the least developed countries in identifying and addressing their urgent adaptation needs. Case studies of NAPA implementation, such as in Bangladesh and Malawi, highlight the importance of integrating equity considerations into adaptation planning and implementation to ensure that vulnerable populations are prioritized.

Adaptation and Vulnerability in The UNFCCC is a critical component of collective climate action. In its 1992 Earth Summit, the UNFCCC established five principles that included "the precautionary principle," "prevention of dangerous interference with the climate system," and "adaptive measures to cope with adverse effects" (Kirchhoff et al., 2019). This recognition of adaptation as an essential factor for responding to global warming has been reaffirmed by subsequent international agreements such as the Paris Agreement. Despite this progress, challenges remain due to inadequate implementation and scientific knowledge gaps concerning how environmental changes will manifest at various scales (Mendelsohn & Schlesinger, 2018). Additionally, there are increasing complexities associated with assessing impacts across multiple sectors including food production systems which can be subject to significant variability from year-to-year (Desjardins et al., 2020). Further complicating matters is understanding how varying levels of vulnerability among different populations intersects with adaptation strategies for greater effectiveness (Buchy et al., 2018; Kapsheev & Hynes, 2018).

The need for effective networked governance accessible across all stakeholders was recognized at COP 21 when parties adopted Agenda 2030 Sustainable Development Goals 16 through 19 which emphasize peacebuilding and inclusive decision-making processes based on access to reliable data sources and information sharing technologies (Schellnhuber et al., 2017). This emphasis reflects growing awareness that traditional top-down approaches have proven ineffective due largely to lack of trust between leaders who often come from vastly different cultural contexts (Atalan Helicke, 2019; Bailey et al., 2012; Bernstein & Hoffmann, 2018). Adopting a more collaborative approach towards problem solving allows policy makers greater flexibility while allowing local communities direct input into mitigation efforts or actions taken post disaster thereby creating better outcomes overall. Furthermore, it encourages new paths forward addressing cross cutting issues such as poverty alleviation in tandem with sustainable development goals related specifically to reducing emissions including renewable energy infrastructure investments or transitioning away from carbon intensive practices like deforestation (Liu et al., 2019; Xuetal. 2021).

The UNFCCC emphasizes the importance of technology transfer and capacity building to support developing countries in their climate action efforts (Cody, 2018; Dey et al., 2017; Mar et al., 2023; Watabe & Gilby, 2020; Wyborn, 2015; Yates, 2012). Equity and justice considerations are central to these efforts, ensuring that developing countries have access to environmentally sound technologies and the necessary capacity to implement climate projects. For instance, the Technology Mechanism established under the UNFCCC supports technology transfer through the Technology Executive Committee (TEC) and the Climate Technology Centre and Network (CTCN). Scholarly work by Gupta et al. (2017) emphasizes the importance of equity in technology transfer, arguing that it should consider the specific needs and priorities of developing countries and promote the transfer of both mitigation and adaptation technologies. Technology transfer is a key component of collective climate action, as it enables countries to share knowledge and resources in order to reduce emissions. According to the United Nations UNFCCC, technology transfer involves the dissemination of technologies, know-how, processes or methods among countries for reducing their greenhouse gas emissions or enhancing their capacity to adapt to climate change (United Nations Framework Convention on Climate Change [UNFCCC], 2018). It also references financing considerations (Figure 7.8).

Figure 7.8. Three main application areas for technology innovation in the climate realm.

Source: UNEP-CCC.

Capacity building has emerged as an important tool within UNFCCC's framework for collectively addressing global warming through networked governance (Fidelis et al., 2019; Heikkila & Gerlak, 2019; Neville et al., 2019; Vasconcelos et al., 2015; Wernli et al., 2023). It focuses on improving access to technical information, training programs and other forms of assistance which can help create equitable opportunities in both industrialized and developing countries with regard to developing green energy solutions (Vakhratyan & Schreurs, 2020). In this way improved capacities allow governments at all levels – international organizations such as UNFCCC; national ministries;

local authorities; non-state actors including civil society groups – have all engaged in capacity building initiatives under the auspices of UNFCCC's networked climate governance. Capacity building activities are essential in providing support for effective implementation of policies aimed at tackling climate change by strengthening existing institutions involved in mitigating its effects while creating new ones when needed (Ribeiro et al., 2019). This helps ensure that these measures are carried out efficiently during times where financial support may be limited. Without adequate capacity building efforts any potential progress gained from technological advancements would not be able to take effect due lack of proper institutional structure necessary for successful policy implementation. Therefore, Financial Support should go hand-in-hand with technology transfers and capacity building if we hope to make tangible changes towards achieving our goals set forth by collective climate action initiatives like those enacted under UNFCCCs purview.

Equity and justice are integral to financial support mechanisms established under the UNFCCC. The principle of "new and additional" financial resources highlights the need for developed countries to provide adequate and predictable financial support to developing countries for their climate actions (Koontz et al., 2015; Krause et al., 2021; Laycock & Mitchell, 2019; Lee et al., 2014). The GCF, established under the UNFCCC, is a significant financial mechanism aimed at mobilizing resources to support developing countries in their climate efforts. The GCF emphasizes equity and justice by providing funding for both mitigation and adaptation projects, prioritizing the most vulnerable countries and communities. When discussing the financial support of UNFCCC under collective climate action and networked climate governance, it is important to consider the Principle of Common but Differentiated Responsibilities and Respective Capabilities (CBDR RC). This principle dates back to 1992 when participating parties in the UNFCCC agreed that developed countries should take greater responsibility for mitigating global warming due to their higher levels of emissions over developing countries. As such, financial support from these nations has become an integral aspect within collective climate action initiatives.

Financial assistance from developed countries towards less affluent areas can provide them with resources necessary for mitigation measures (Davies et al., 2018; Deka & Goswami, 2022; Dey et al., 2017; Heikkila & Gerlak, 2019; Tosun & Schoenefeld, 2017). However, this process requires a delicate balance between equitable distribution and nation-specific factors (Kamimura et al., 2018). For example, Kamimura et al. propose that contributions made by donors must be tailored according to each country's specific requirements so as not only adhere to CBDR RC but also guarantee effective implementation. In addition, they suggest that donor states should collaborate closely with recipient nations throughout all stages of policy development while ensuring mutual accountability systems are in place. Accordingly, recent data indicates that Global South regions receive 90% or more of international public finance devoted solely towards adaptation activities meant to reduce vulnerability due both natural disasters and long-term effects associated with climate change (Hochrainer-Stigler & Huggel 2019). Such investments have been crucial for helping communities adapt through various channels including insurance programs against extreme weather events (Lemke & Renn 2020); infrastructure projects designed to limit flooding damages (Nguyen et al., 2017); and sustainable agriculture practices aiming at reducing erosion (Banse & Hoppe 2016). While much progress has been made since CBDR RC was first established in 1992, there is still room for improvement particularly concerning how funds are directed across different sectors and geographical locations.

7.5. THE PRINCIPLE OF COMMON BUT DIFFERENTIATED RESPONSIBILITIES AND RESPECTIVE CAPABILITIES (CBDR-RC)

The Principle of CBDR-RC is a fundamental concept within the framework of climate governance (Neville et al., 2019; Niles & Lubell, 2012; Niles et al., 2021; O'Brien et al., 2021; Olivier & Berardo, 2022). It recognizes that while addressing climate change is a shared responsibility, countries have varying historical contributions to greenhouse gas emissions and diverse capacities to address the challenges. CBDR-RC seeks to ensure equity and justice by acknowledging these differences and promoting fair distribution of burdens and benefits in collective climate action. This section will explore the significance of CBDR-RC in achieving equity and justice within the wider framework of collective climate action and networked climate governance. The principle of CBDR-RC was first established in the UNFCCC in 1992. It recognizes that all countries share a common responsibility to address climate change but emphasizes that developed countries bear a greater responsibility due to their historical emissions and higher capacities. This principle acknowledges the historical and current disparities in carbon emissions, economic development, and technological capabilities among nations.

Collectively, the concept of 'CBDR-RC,' as stipulated in Article 3.1 of the UNFCCC 1992 document is essential when seeking equity and justice under collective climate action. The notion was developed to address how countries should respond differently depending upon their capacity levels or level of contribution to global emissions while taking into account "special situations" such as small island nations that are especially vulnerable to the impacts of climate change (Tian & Jie, 2019). In essence, CBDR-RC compels those who have contributed most heavily to greenhouse gas emissions historically – usually industrialized countries – to bear a greater responsibility for combating current global warming trends than less industrialized states with lower emission rates; something which has been confirmed by various studies over time including one conducted by van Asselt et al., (2018).

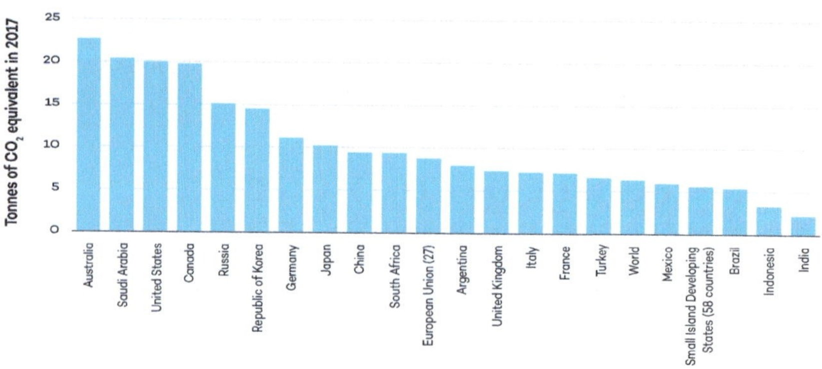

Figure 7.9. G20 members' and vulnerable countries' per capita GHG emissions (excl. land use change and forestry).

Source: Luomi (2020) via International Institute for Sustainable Development (IISD).

According to this research paper published in Global Environmental Politics last year, "the principle reflects an understanding that historical responsibilities play a role in determining present obligations;" thus emphasizing why it remains so important today particularly since many developing countries still lack access to adequate resources with which they can combat these issues adequately. However, although there have been several advances made within areas such as technology transfer initiatives and financial aid programs intended designed specifically help emerging economies tackle their own unique challenges associated with climate change adaptation measures (Carmin & Olmstead 2010), much more needs be done if true equity amongst all states is ever going achieve. In fact according WRI report from 2017 entitled 'Global Efforts To Mitigate And Adapt To Climate Change: A Progress Report,' only 4% of total public funds dedicated towards tackling climate related problems globally were allocated directly towards helping least developing nations even though collectively they accounted for nearly 20% world`s population at risk due natural disasters during same period (WRI 2017). This stark discrepancy clearly highlights just how far we still need to go before any real progress can be achieved in terms of achieving equitable outcomes between different nation's ability to fight against climatic changes caused by human activities (Figure 7.9).

CBDR-RC plays a vital role in achieving equity and justice in collective climate action by promoting fairness and inclusivity (Fried et al., 2022; Gore, 2010; Harmon et al., 2023; Niles & Lubell, 2012; Rudnick et al., 2019; Vantaggiato et al., 2023). It recognizes that developing countries, despite contributing less to historical emissions, often bear a disproportionate burden in terms of climate impacts and adaptation costs. By emphasizing differentiation, CBDR-RC helps ensure that climate actions are not unfairly imposed on developing nations, potentially hindering their economic and social development. CBDR-RC enables the provision of financial and technological support from developed to developing countries. This support assists developing nations in implementing climate mitigation and adaptation measures, fostering sustainable development pathways. Through financial mechanisms such as the GCF, developed countries contribute to addressing the financial barriers faced by developing nations in tackling climate change. This financial support helps level the playing field and facilitates equitable participation in collective climate action. Additionally, CBDR-RC recognizes the importance of technology transfer and capacity building. Developed countries possess advanced technological capacities and have a responsibility to share these resources with developing countries. By facilitating the transfer of clean and sustainable technologies, CBDR-RC promotes the ability of developing countries to address climate change effectively. This technology transfer empowers nations to transition to low-carbon economies, enhancing their resilience and reducing their vulnerability to climate impacts.

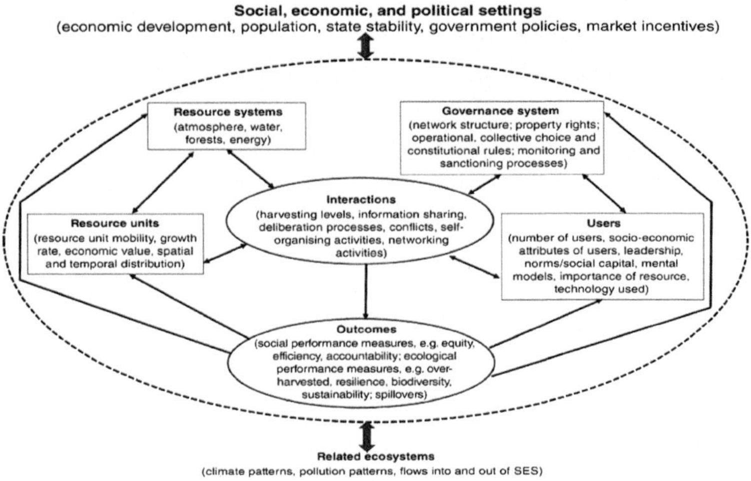

Figure 7.10. Contextual considerations when implementing networked climate governance.

Source: Zia (2021).

CBDR-RC is particularly relevant in the context of networked climate governance (Figure 7.10), where multiple actors, including national governments, subnational entities, civil society organizations, and private sectors, collaborate to address climate change. The principle helps guide decision-making processes, ensuring that the voices and concerns of all stakeholders, particularly those from vulnerable and marginalized communities, are considered (Calliari et al., 2019; Fidelis et al., 2019; Gunningham, 2012; Koefoed, 2019; Perz et al., 2022; Sultana et al., 2019). Within networked climate governance, CBDR-RC promotes inclusivity by recognizing and valuing diverse perspectives, needs, and capabilities. It encourages dialog and collaboration between countries with varying levels of development, allowing for the exchange of knowledge, experiences, and best practices. This collaborative approach fosters mutual learning and promotes the development of context-specific solutions to climate challenges, considering the unique socio-economic and environmental contexts of different countries. CBDR-RC also plays a critical role in promoting accountability within networked climate governance. It emphasizes the need for transparency in reporting emissions, implementing climate actions, and providing support to developing countries. This transparency allows for monitoring and assessment of progress and ensures that commitments made by countries are fulfilled. By holding countries accountable to their respective responsibilities and capabilities, CBDR-RC strengthens the overall effectiveness of collective climate action. By recognizing historical emissions, economic disparities, and varying capacities among countries, CBDR-RC promotes fairness in burden-sharing and facilitates the provision of financial and technological support to developing nations. Within networked climate governance, CBDR-RC fosters inclusivity, promotes accountability, and encourages collaboration across diverse stakeholders. As we navigate the challenges of climate change, CBDR-RC remains crucial in ensuring a just and equitable transition to a sustainable future for all.

Networked climate governance is an essential component of collective climate action aimed at seeking equity and justice. Developing countries often lack the capacity and resources to manage their own environmental projects, so they must seek outside sources for support (Lahari & Kumar, 2018). Specifically, networked climate governance involves multiple actors—including non-governmental organizations (NGOs), private companies, governments, local communities—collaborating on shared efforts to combat global warming through mitigation strategies such as emission reductions or adaptation initiatives like drought resilience programs (Kim et al., 2020). These networks can be organized in different ways: horizontally across sectors; vertically between international institutions and national/local entities; or in a hybrid model that includes both approaches (Chaturvedi et al., 2019). They may also involve public-private partnerships (PPPs) with businesses taking responsibility for some of the costs associated with achieving sustainability goals. One example of this type of collaboration comes from India's National Action Plan on Climate Change initiative which was launched by Prime Minister Manmohan Singh in 2008. This plan requires all stakeholders including state governments, NGOs and citizens groups to work together toward a common goal: mitigating greenhouse gas emissions by 20–25% below 2005 levels by 2025 (Mahapatra & Kundu 2013). The program has seen some successes due to its collaborative nature – it has significantly reduced air pollution within Indian cities since implementation began according calculations made using satellite data provided by NASA's Aura Mission Ozone Monitoring Instrument – but there have been challenges too – many states are still struggling financially when trying to implement new energy efficiency measures required under the plan (Goswami et al., 2017). Nevertheless, it demonstrates how concerted effort among different actors can achieve significant results towards tackling pressing environmental issues like global warming.

CHAPTER 7 SUMMARY

In conclusion, the pursuit of equity and justice in climate action and governance is imperative for addressing the complex challenges of climate change. This chapter has explored the principles and mechanisms that underpin equity and justice within the wider framework of collective climate action and networked climate governance. We have seen how the Principle of CBDR-RC plays a pivotal role in ensuring fairness, inclusivity, and accountability in climate governance. The principle of CBDR-RC recognizes historical disparities, economic differences, and varying capacities among nations, emphasizing the need for differentiated responsibilities and support mechanisms. It highlights the importance of financial assistance, technology transfer, and capacity building to enable developing countries to actively participate in climate mitigation and adaptation efforts. By promoting fairness in burden-sharing and facilitating collaboration between developed and developing countries, CBDR-RC contributes to the achievement of equitable outcomes in collective climate action. Moreover, this chapter has highlighted the significance of networked climate governance in fostering equity and justice. The involvement of diverse stakeholders, such as national and subnational governments, civil society organizations, and the private sector, is essential for a comprehensive and inclusive approach to climate change. Networked governance allows for the integration of diverse perspectives, knowledge, and experiences, ensuring that the voices of marginalized and vulnerable communities are heard and taken into account.

The next and final chapter of this book, titled "Intersectionality in Climate Change Action and Networked Governance," will build upon the foundation laid in this chapter by examining the intersectionality of climate change impacts and responses. It will explore how climate change intersects with other forms of social and environmental inequality, such as gender, race, class, and indigenous rights. The chapter will delve into the ways in which climate action and networked governance can address these intersecting dimensions of injustice and ensure that climate policies and initiatives are sensitive to the needs and experiences of diverse populations. By exploring intersectionality in climate change action and networked governance, the final chapter will contribute to a deeper understanding of the complex and interconnected nature of climate justice. It will provide insights into how climate governance can be more inclusive, equitable, and effective in addressing the multifaceted challenges of climate change. Through an examination of intersectional perspectives, the chapter will offer practical recommendations and policy implications for enhancing the equity and justice aspects of collective climate action. Generally, achieving equity and justice in climate action and governance is essential for a sustainable and resilient future. The principles of equity and justice, as exemplified by CBDR-RC and networked climate governance, provide the foundation for fair and inclusive decision-making processes and ensure that the voices of all stakeholders are heard. The next chapter on intersectionality in climate change action and networked governance will further deepen our understanding of the complexities of climate justice and contribute to the development of more comprehensive and transformative climate policies and initiatives. Together, these chapters provide a holistic view of the challenges and opportunities in addressing climate change in a just and equitable manner.

CHAPTER 8

Intersectionality in Climate Change Action and Networked Governance

CONTENTS

The final chapter of this book, titled "Intersectionality in Climate Change Action and Networked Governance," explores the interconnected nature of climate change impacts and responses. Climate change is not just an environmental issue; it intersects with various dimensions of social inequality, including gender, race, class, and indigenous rights. Understanding these intersections is crucial for achieving equitable and just outcomes in collective climate action and networked climate governance. Intersectionality acknowledges that individuals and communities experience multiple forms of discrimination and disadvantage simultaneously. It recognizes that climate change impacts and vulnerabilities are shaped by intersecting social identities and power dynamics. For example, women in marginalized communities may face distinct challenges in adapting to climate change due to their gender roles, limited access to resources, and social marginalization. Indigenous communities may face disproportionate climate impacts due to their close relationship with and dependence on the natural environment. This chapter delves into how climate change action and networked governance can address these intersecting dimensions of injustice. By incorporating an intersectional lens, we can develop more nuanced and inclusive approaches to climate policies, programs, and decision-making processes. Recognizing and addressing the specific needs and experiences of different groups is essential for promoting equity, justice, and resilience in the face of climate change.

The chapter begins by providing a theoretical foundation on intersectionality and its relevance to climate change. It explores how intersecting identities and power structures shape vulnerability, adaptation, and mitigation strategies. By understanding the complex ways in which social inequalities intersect with climate change, we can identify and address the root causes of injustice and inequality. Furthermore, the chapter examines the role of networked governance in integrating intersectionality into climate action. Networked governance brings together diverse stakeholders from different sectors and levels of governance, fostering collaboration, knowledge exchange, and collective decision-making. By incorporating intersectional perspectives within networked governance structures, we can ensure that the voices and experiences of marginalized communities are heard and inform policy and action. The chapter also highlights practical examples of intersectional climate action and networked governance initiatives from around the world. It explores case studies where intersectional approaches have been integrated into policy frameworks, community-based projects, and international agreements. These examples illustrate the potential for transformative change when intersectionality is embraced in climate governance.

In addition, the chapter explores the challenges and opportunities associated with integrating intersectionality into climate change action. It addresses issues such as data gaps, power dynamics, and institutional barriers that may hinder the full integration of intersectional perspectives. It also discusses strategies and recommendations for policymakers, practitioners, and researchers to overcome these challenges and promote more inclusive and equitable climate action. Ultimately, the goal of this final chapter is to provide a comprehensive understanding of how intersectionality can enhance collective climate action and networked governance. By recognizing and addressing the interconnected nature of social inequalities and climate change, we can develop more effective and just strategies to tackle the climate crisis. Through the integration of intersectional perspectives, we can create climate policies and initiatives that are sensitive to the diverse needs and experiences of individuals and communities, promoting resilience, equity, and justice in the face of a changing climate.

8.1. THE REALITY OF ENVIRONMENTAL RACISM

Using an intersectionality lens, the examination of environmental racism and its policy implications in collaborative climate action and networked climate governance reveals the disproportionate burden borne by marginalized communities and the urgent need for equitable solutions (Bernstein & Hoffmann, 2018; Bodin, 2017; Herzog & Ingold, 2019; Luthe & Wyss, 2016; Willems & Giezen, 2022). Environmental racism refers to the intersection of racial discrimination and environmental degradation, where minority and marginalized communities face higher levels of pollution, hazardous waste, and limited access to resources necessary for adaptation and resilience in the face of climate change.

This discussion will critically analyze the realities of environmental racism and explore its policy implications within the context of collective climate action and networked climate governance. Intersectionality recognizes that race intersects with other social identities, such as class, gender, and geography, shaping experiences of vulnerability and resilience in the face of environmental challenges. Historically, marginalized communities, particularly communities of color, have been disproportionately exposed to environmental hazards, including toxic waste sites, industrial pollution, and degraded natural resources (Figure 8.1). This systemic injustice exacerbates the impacts of climate change on these communities, perpetuating existing social inequalities.

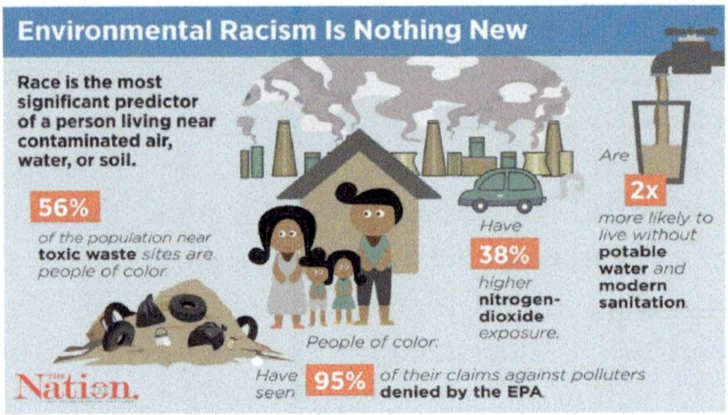

Figure 8.1. Linking race to environmental vulnerability and consequences.

Source: Covert (2016) via The Nation.

One key aspect of addressing environmental racism in climate action is through policy interventions that prioritize environmental justice and equity (van Bers et al., 2019; van Buuren & Eshuis, 2009; Van Popering-Verkerk et al., 2022). Policies should aim to rectify the historical and ongoing injustices faced by marginalized communities, ensuring that climate adaptation and mitigation efforts are targeted towards those most affected. This requires an intersectional approach that takes into account the unique needs, perspectives, and experiences of different racial and ethnic groups. In networked climate governance, collaboration among diverse stakeholders is essential for addressing environmental racism effectively. This collaboration should include representatives from affected communities, grassroots organizations, and community leaders who can contribute their expertise and advocate for the interests of marginalized groups. Networked governance structures should promote inclusive decision-making processes, enabling meaningful participation of affected communities in shaping policies and initiatives. Furthermore, policy frameworks should prioritize community-based and participatory approaches to climate action. This means empowering affected communities to actively participate in decision-making processes, providing them with resources, capacity-building support, and access to information. By engaging communities in the design and implementation of climate initiatives, policy frameworks can ensure that the solutions address the specific needs and priorities of marginalized groups.

In terms of climate finance, policy implications should involve the equitable distribution of resources to address environmental racism (Willems

& Giezen, 2022; Wyborn, 2015; Yates, 2012). Financial mechanisms such as the Green Climate Fund (GCF) and climate finance commitments from developed countries should prioritize funding for adaptation and mitigation projects in marginalized communities, specifically targeting areas affected by environmental racism. This financial support should be accompanied by capacity-building initiatives that enable communities to effectively utilize resources and implement sustainable climate solutions. Furthermore, policies should emphasize the integration of environmental justice considerations into environmental impact assessments and decision-making processes. This would involve conducting thorough assessments of the potential environmental and social impacts of projects on marginalized communities and ensuring that the voices and concerns of these communities are heard and given due consideration in decision-making.

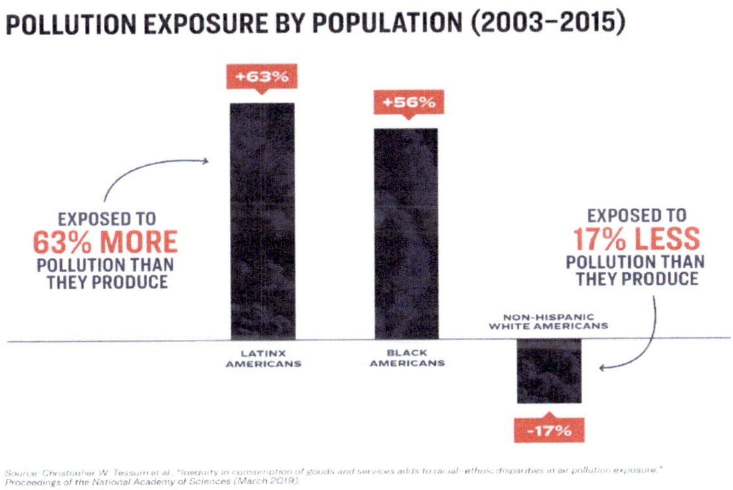

Figure 8.2. Racialized exposure to pollution in the USA between 2003 and 2015.

Source: Beech (2020) via The World Economic Forum.

To address environmental racism in collaborative climate action, it is crucial to recognize and dismantle the systemic structures that perpetuate these inequities (Fisher & Nasrin, 2021; Fitch-Roy et al., 2020; Fried et al., 2022; Friedman et al., 2020). This includes addressing underlying issues of racism, discrimination, and power imbalances that contribute to environmental injustices (Figure 8.2 for an example of how a common climate issue such as exposure to pollution is racialized). Policies should promote inclusive and anti-racist approaches to climate governance, aiming to create transformative change that challenges the status quo and centers the voices of marginalized communities. Overall, addressing environmental racism within the context of collective climate action and networked climate governance requires an intersectional lens that acknowledges the unique experiences and vulnerabilities of marginalized communities. Policy implications should prioritize environmental justice, community participation, equitable distribution of resources, and anti-racist approaches. By integrating these principles into climate policies and initiatives, we can strive for a more equitable and just transition towards a sustainable future for all.

In recent years, there has been increasing recognition of the disproportionate effects of climate change on certain populations. Environmental racism is a form of discrimination that disproportionately affects communities and individuals due to their race or ethnicity (Robertson & Richards-Doran, 2020). In terms of collective climate action and networked climate governance, environmental racism manifests itself in inequitable access to resources such as green technology innovation and information about health risks associated with air pollution (Merrill et al., 2018; Chang et al., 2019). This lack of access can create an unequal burden for minority groups which further perpetuates systemic inequality. Additionally, women are particularly vulnerable to the impacts from global warming given traditional gender roles related to land use and resource management (Borras et al., 2018; Sultana & Yossofzai, 2019). As they are often responsible for providing food security within households while also caring for children with limited financial support available, women face unique challenges when it comes to dealing with extreme weather events linked to rising temperatures. Ultimately, mitigating environmental racism requires an intersectional approach that takes into account both social justice concerns as well as the need for equitable access to resources necessary for adapting successfully to changing climatic conditions.

8.2. UNDERSTANDING THE ROLE OF WOMEN IN CLIMATE CHANGE

Using an intersectionality lens, it is crucial to critically discuss the role of women in climate change and its policy implications within the framework of collaborative climate action and networked climate governance (Partelow et al., 2020; Pavanelli et al., 2023; Perz et al., 2022). Intersectionality recognizes that women's experiences and vulnerabilities are shaped by the intersection of gender with other social identities such as race, class, ethnicity, and geography. This perspective sheds light on the unique challenges faced by women in climate change contexts and underscores the need for gender-responsive policies and inclusive decision-making processes. Women play a significant role in climate change adaptation, mitigation, and resilience efforts. They are often at the forefront of community-based initiatives, as they bear the brunt of climate impacts and are essential in safeguarding the well-being of their families and communities. However, systemic gender inequalities, coupled with intersecting forms of discrimination, create barriers that hinder women's active participation and limit their influence in climate decision-making. One key aspect of addressing gender disparities in climate action is through policy interventions that promote gender equality and empower women (Figure 8.3). Policies should aim to address the underlying gender inequalities that intersect with climate change, including unequal access to resources, limited decision-making power, and discriminatory cultural norms. This involves ensuring that women have equal access to education, training, and financial resources to actively engage in climate-related activities.

Networked climate governance provides an opportunity to integrate gender perspectives into decision-making processes and policy development (Chu, 2018; Cody, 2018; de Macedo & Jacobi, 2019; Fitch-Roy et al., 2020; Orach & Schluter, 2016; Zhan, 2022). Collaborative platforms should incorporate diverse voices, including women's representation, to ensure that gender-specific concerns are integrated into climate initiatives. Women's organizations and gender-focused networks should be actively engaged in these networks to ensure that gender issues are effectively addressed and that the experiences and expertise of women are recognized and valued. Moreover, policy frameworks should integrate gender-responsive approaches to climate finance and resource allocation. This includes providing financial mechanisms that prioritize women's access to funding and resources for climate adaptation and mitigation projects. Additionally, policies should ensure that gender considerations are mainstreamed into climate finance mechanisms such as the GCF, making funding more accessible to women-led initiatives and projects that address gender-specific climate vulnerabilities.

In terms of capacity-building, policy implications should focus on enhancing women's leadership and decision-making capacities in climate governance (Gore, 2010; Hitziger et al., 2019; Kim et al., 2019; O'Brien et al., 2021; Perz et al., 2022). This involves providing training, mentorship, and opportunities for women to actively engage in climate policy and decision-making processes at all levels.

Figure 8.3. Harnessing women in the management of climate change.

Source: Women Deliver.

By fostering women's leadership and participation, policy frameworks can benefit from their unique perspectives and experiences, resulting in more inclusive and effective climate action. Addressing gender-based violence and ensuring women's safety in climate action is another critical policy implication. Climate change exacerbates existing vulnerabilities and can increase the risks of violence against women, particularly in disaster-prone areas. Policies should prioritize measures to prevent and respond to gender-based violence in climate-affected communities, ensuring that women's safety and well-being are protected.

Policy frameworks should recognize and value the unpaid care and domestic work performed by women, which is often intensified in the context of climate change (Hale & Roger, 2014; Hoffmann & Villamayor-Tomas, 2023; Krause et al., 2021; Tuda & Machumu, 2019). By acknowledging the gendered division of labor and promoting policies that alleviate the burden of unpaid care work, such as providing access to clean energy and infrastructure, women can have more time and resources to engage in climate-related activities and decision-making. Generally, adopting an intersectionality lens to examine the role of women in climate change highlights the importance of gender-responsive policies and inclusive decision-making processes. It underscores the need to address systemic gender inequalities and the intersecting forms of discrimination that shape women's experiences and vulnerabilities in the face of climate change. By integrating gender perspectives into collaborative climate action and networked climate governance, policies can empower women, recognize their expertise, and create a more equitable and effective response to the challenges of climate change.

Intersectionality is a concept that focuses on the multiple, intersecting identities of individuals and how these shape their experiences (Den Uyl & Russel, 2018; Gordon, 2016; Morrison et al., 2017; Sayegh, 2020; Yoder et al., 2022). This framework has been increasingly used to analyze climate change impacts on different groups in society including women who face unique challenges due to gender-based discrimination (Berrin, 2020). The effects of climate change are experienced differently amongst various social classes, race/ethnicity and genders and this can have profound implications for collective action towards adaptation and mitigation strategies (Gentry & Eckerberg, 2018). Women often bear an unequal burden when it comes to natural resource management as they do not possess equal access or influence over decision making processes (Folke et al., 2019). Hence, the intervention by the fifth Sustainable Development Goal (Figure 8.4).

In order to address this imbalance, new forms of governance such as networked climate governance need to be implemented where all stakeholders are included equally regardless of gender. According to UNEP's Gender Action Plan II which aims at promoting gender equality by "empowering women through knowledge building initiatives" (UN Environment Program [UNEP], 2017), there needs to be more focus placed on the role females play in responding effectively with regards sustainable development goals set out by the Paris Climate Agreement. Indigenous people also make key contributions into global efforts towards adapting and mitigating against climate change although much attention still needs be given specifically within local contexts regarding land rights issues concerning traditional lands held collectively by communities (Langston & Anderson 2016). Therefore, incorporating indigenous knowledge systems into existing frameworks for networked climate governance could prove beneficial when designing equitable solutions around natural resources management thus enabling both men and women alike full participation in decision making processes related directly affecting them.

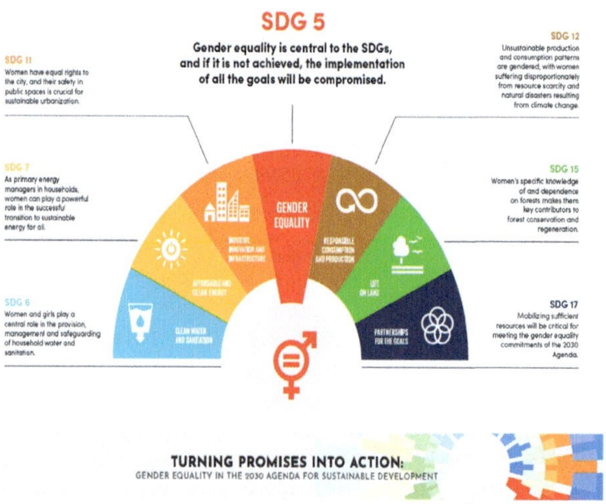

Figure 8.4. The role SDG5 in achieving all SDGs.

Source: Sinha (2019) via the world economic forum.

8.3. INDIGENOUS KNOWLEDGE AND NATURAL RESOURCE MANAGEMENT

Using an intersectionality lens, it is essential to critically discuss Indigenous knowledge and natural resource management and its policy implications within the context of collaborative climate action and networked climate governance (Bowen et al., 2017; Brandenberger et al., 2022; Brewer, 2013; Brewer et al., 2017; Buchtmann et al., 2022). Intersectionality recognizes the complex interplay between Indigenous identity, culture, land, and the impacts of climate change. Indigenous communities have unique knowledge systems and traditional practices that have sustained their relationships with the environment for generations. Understanding and integrating Indigenous knowledge into climate policies and governance frameworks is crucial for fostering resilience, promoting environmental sustainability, and upholding Indigenous rights. Indigenous knowledge systems are deeply rooted in the interconnectedness of ecosystems, emphasizing the holistic understanding of the environment and the interdependence of all living beings. Indigenous communities possess valuable insights on sustainable resource management, biodiversity conservation, and climate adaptation strategies. However, historical and ongoing colonial practices have often marginalized Indigenous knowledge, resulting in the marginalization of Indigenous peoples and the degradation of their lands.

One key aspect of collaborative climate action and networked climate governance is the recognition and respect for Indigenous rights and self-determination (Hale & Roger, 2014; Hamilton & Lubell, 2018; Harmon et al., 2023; Hawkins et al., 2018). Policies should prioritize the free, prior, and informed consent of Indigenous communities in all decision-making processes that affect their lands, resources, and traditional knowledge. This includes engaging Indigenous representatives and organizations in collaborative platforms and policy dialogs, ensuring that their voices are heard and their perspectives inform climate initiatives. Moreover, policy frameworks should actively incorporate and value Indigenous knowledge in natural resource management and climate adaptation strategies. Traditional ecological knowledge offers insights into climate patterns, biodiversity conservation, and sustainable land management practices. By integrating Indigenous knowledge into climate policies, governments can enhance the effectiveness and resilience of adaptation and mitigation measures. This requires creating space for intercultural dialogs, knowledge-sharing, and co-production of knowledge between Indigenous communities, scientists, and policymakers.

Networked climate governance can play a vital role in facilitating the inclusion of Indigenous knowledge. Collaborative platforms should create opportunities for Indigenous peoples to share their knowledge, experiences, and best practices, enabling cross-cultural learning and the co-design of climate solutions (Virah-Sawmy et al., 2019; Voogd et al., 2022; Watabe & Gilby, 2020; Wernli et al., 2023). Additionally, networked governance structures should provide support for capacity-building initiatives that strengthen Indigenous communities' participation in decision-making processes and equip them with the necessary skills and resources to engage effectively. In terms of land and resource governance, policy implications should prioritize the recognition and protection of Indigenous lands, territories, and resources. Indigenous peoples have been stewards of their lands for generations, and their rights to land and resources should be respected and upheld. Policies should recognize Indigenous land tenure systems, support land titling processes, and provide legal mechanisms for Indigenous communities to assert control over their territories. Additionally, policies should address the issue of land dispossession and support efforts to restore and reclaim Indigenous lands and territories.

Schwartz's Humanitarian Values	
1. Equality (equal opportunity for all)	29. A world of beauty (beauty of nature and the arts)
2. Inner harmony (at peace with myself)	30. Social justice (correcting injustice, care for the weak)
3. Social power (control over others, dominance)	31. Independent (self-reliant, self-sufficient)
4. Pleasure (gratification of desires)	32. Moderate (avoiding extremes of feeling and action)
5. Freedom (freedom of action and thought)	33. Loyal (faithful to my friends, group)
6. A spiritual life (emphasis on spiritual not material matters)	34. Ambitious (hard-working, aspiring)
7. Sense of belonging (feeling that others care about me)	35. Broadminded (tolerant of different ideas and beliefs)
8. Social order (stability of society)	36. Humble (modest, self-effacing)
9. An exciting life (stimulating experiences)	37. Daring (seeking adventure, risk)
10. Meaning in life (a purpose in life)	38. Protecting the environment (preserving nature)
11. Politeness (courtesy, good manners)	39. Influential (having an impact on people and events)
12. Wealth (material possessions, money)	40. Honoring of parents and elders (showing respect)
13. National security (protection of my nation from enemies)	41. Choosing own goals (selecting own purposes)
14. Self-respect (belief in one's own worth)	42. Healthy (not being sick physically or mentally)
15. Reciprocation of favors (avoidance of indebtedness)	43. Capable (competent, effective, efficient)
16. Creativity (uniqueness, imagination)	44. Accepting my portion in life (submitting to life's circumstances)
17. A world at peace (free of war and conflict)	45. Honest (genuine, sincere)
18. Respect for tradition (preservation of time-honored customs)	46. Preserving my public image (protecting my "face")
19. Mature love (deep emotional and spiritual intimacy)	47. Obedient (dutiful, meeting obligations)
20. Self-discipline (self-restraint, resistance to temptation)	48. Intelligent (logical, thinking)
21. Detachment (from worldly concerns)	49. Helpful (working for the welfare of others)
22. Family security (safety for loved ones)	50. Enjoying life (enjoying food, sex, leisure, etc.)
23. Social recognition (respect, approval by others)	51. Devout (holding to religious faith and belief)
24. Unity with nature (fitting into nature)	52. Responsible (dependable, reliable)
25. A varied life (filled with challenge, novelty, and change)	53. Curious (interested in everything, exploring)
26. Wisdom (a mature understanding of life)	54. Forgiving (willing to pardon others)
27. Authority (the right to lead or command)	55. Successful (achieving goals)
28. True friendship (close, supportive friends)	56. Clean (neat, tidy)

Figure 8.5. Examples of humanitarian values that could inform indigenous knowledge.

Source: Struch & Schwartz (2002) as cited in Gratani (2015).

Furthermore, policy frameworks should promote the revitalization and preservation of Indigenous languages, cultures, and traditional practices (Figure 8.5 for humanitarian values that could inform this process). Cultural revitalization is closely intertwined with climate resilience, as Indigenous cultures and languages are often inseparable from their knowledge of the environment (Kelman & Clark-Ginsberg, 2022; Kim et al., 2019; Kithiia, 2015; Koefoed, 2019). Supporting Indigenous cultural resurgence contributes to the preservation of traditional ecological knowledge and strengthens Indigenous communities' adaptive capacities in the face of climate change. Moreover, policies should address the socioeconomic disparities faced by Indigenous communities, which intersect with climate vulnerability. Adequate and targeted resources should be allocated to support sustainable economic development initiatives in Indigenous territories, ensuring that communities have the capacity to adapt to climate change while maintaining their cultural integrity and traditional livelihoods. Therefore, adopting an intersectionality lens to examine Indigenous knowledge and natural resource management highlights the importance of recognizing and respecting Indigenous rights, integrating Indigenous knowledge systems, and promoting inclusive and equitable governance structures. Policy implications include the recognition of Indigenous self-determination, the integration of Indigenous knowledge in climate policies, the protection of Indigenous lands and resources, and support for cultural revitalization and socioeconomic development. By centering Indigenous voices and knowledge in collaborative climate action and networked climate governance, policies can contribute to environmental sustainability. Moreover, it is important to understand the differences between local and global knowledge (Figure 8.6) so that the former can be centered in climate change responses.

	Local knowledge	Global knowledge
Applicability	Often seen as specific to the local socio-cultural, agro-ecological context	Meant to have universal applicability
Uses	Guides the decisions of peasant farmers, informs interventions via participatory approach	Key determinant of development, notably via schooling, extension services
Validation	Wealth of practical experience based on day-to-day management and observation	Scientific method, technological applications
Transmission	Transmitted orally from fellow farmers or derived from own observations and experience	Transmitted via formal information provision media such as education, extension or radio
Status	May be low	Tends to be high
Costs	Relatively low, involving observing, listening, experimenting w/ competing practices	May be inaccessible to poor farmers due to costs associated with schooling and radios, or to poor quality rural education
Knowledge gaps	Gaps vis-a-vis local NRs not foreseen, with LK viewed as dynamic and well-suited to the local context	Gaps foreseen, with GK secured only where it is accessible, affordable and cost-effective

Figure 8.6. Comparing local and global knowledge in development literature.

Source: Siedenburg (2022).

Indigenous knowledge and natural resource management are integral parts of collective climate action. A review by Tsilimigka, Hirsch Hadorn and Pohl (2018) found that indigenous peoples have a "unique relationship with their local environment" which is based on millennia-old customary laws and practices (p. 2). They identified the need for holistic approaches to environmental governance where 'traditional' ecological knowledge systems can be integrated into existing policy frameworks in order to better address current climatic challenges such as increasing temperatures, extreme weather events, sea level rise, more frequent droughts or floods. This indicates how networked climate governance could benefit from incorporating traditional forms of managing resources in response to changing climates due to its potential efficacy across multiple scales: within communities but also amongst different stakeholders at regional levels. A study by Tompkins et al (2016) revealed evidence that suggests many Indigenous Peoples actually possess an advanced understanding of the impacts of human actions upon nature; they understand linkages between social injustice – rooted in colonialism – poverty and vulnerability towards global warming while actively engaging in mitigation measures through community-based projects like reforestation efforts or Integrated Coastal Zone Management initiatives (ICZM). According to data from World Resources Institute's Land Cover initiative, up until 2018 there were 96 million hectares managed under ICZM programs worldwide showing positive advances brought about by including these concepts into sustainable development strategies globally. Consequently this has proven how effective 'bottom-up' solutions driven mainly by Indigenous people can be when tackling complex issues related to food security and conservation goals particularly when considering diversity along with equity aspects as part of global policies concerning land use regulation (Tsilimigka et al., 2018).

8.4. NOTIONS OF CLIMATE MIGRATION AND THEIR IMPLICATIONS

Using an intersectionality lens, it is crucial to critically discuss climate migration and its policy implications within the context of collaborative climate action and networked climate governance (Lee et al., 2014; Lee & Koski, 2014; Leventon et al., 2017). Climate change is increasingly recognized as a driver of human migration, as individuals and communities are forced to relocate due to environmental degradation, natural disasters, and the impacts of climate change. Understanding the intersecting dynamics of climate migration, including social, economic, and political factors, is essential for developing effective policies that address the needs and protect the rights of climate migrants. Intersectionality acknowledges that climate migration intersects with other forms of social inequality and marginalization, such as gender, race, socioeconomic status, and nationality. These intersecting identities shape individuals' vulnerabilities, experiences, and access to resources throughout the migration process. Climate migrants often face multiple forms of discrimination and may encounter additional challenges and risks during their displacement and resettlement.

One key aspect of addressing climate migration in collaborative climate action and networked climate governance is the development of comprehensive and rights-based policies (Therville et al., 2019; Tosun & Schoenefeld, 2017; Tuda & Machumu, 2019). Policies should ensure the protection of the human rights of climate migrants, including their right to life, safety, livelihoods, and dignity. This requires adopting an intersectional approach that takes into account the specific needs and vulnerabilities of different groups, such as women, children, indigenous communities, and those with disabilities, among others. Collaborative platforms and networked governance structures should prioritize the inclusion of climate migrants and their representatives in decision-making processes. Their voices should be heard and valued in shaping policies that affect their lives and futures. Moreover, policy frameworks should support the empowerment and participation of climate migrants in governance structures, enabling them to advocate for their rights and contribute their knowledge and experiences to policy discussions.

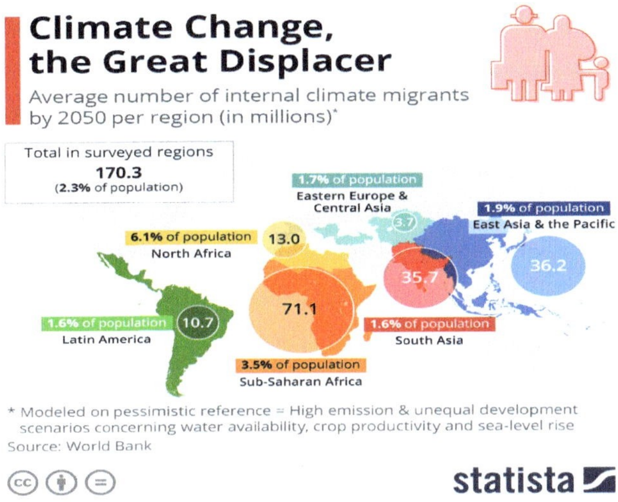

Figure 8.7. The role of climate change in migration patterns.

Source: Statista.

Another critical policy implication is the need to enhance international cooperation and responsibility sharing in addressing climate migration (Blok, 2018; Brandenberger et al., 2022; Laycock & Mitchell, 2019; Sattler & Schroter, 2022; Van Popering-Verkerk et al., 2022). Climate change is a global challenge that requires collective action and shared responsibility. Policies should emphasize burden-sharing and the fair distribution of responsibilities among countries, taking into account historical emissions, capacity, and resources. This includes providing financial and technical support to countries that are most affected by climate migration and lack the capacity to respond effectively. Furthermore, policies should address the root causes of climate migration, including climate change itself, as well as social, economic, and political factors that exacerbate vulnerability and displacement. Mitigation efforts to reduce greenhouse gas emissions should be prioritized to minimize the impacts of climate change and prevent future displacement. Additionally, policies should address social and economic inequalities that contribute to migration by promoting sustainable development, poverty alleviation, and inclusive economic opportunities in climate-affected regions. These interventions are pertinent given the role of climate change in migration patterns (Figure 8.7).

In terms of adaptation, policies should support climate-resilient livelihoods and provide assistance for the integration and resettlement of climate migrants (Bowen et al., 2017; Butler et al., 2022; Fisher & Nasrin, 2021; Homobono et al., 2022; O'Brien et al., 2021). This involves creating opportunities for education, skills training, and access to decent work and social protection. It also requires strengthening social support systems and providing access to basic services, including healthcare, housing, and clean water, in both host communities and areas of origin. Moreover, policies should adopt a gender-responsive approach to climate migration, recognizing and addressing the distinct experiences and vulnerabilities faced by women, who often bear the brunt of climate impacts and displacement. Gender-responsive policies should address the specific needs of women, such as access to reproductive health services, protection from gender-based violence, and opportunities for economic empowerment. Generally, adopting an intersectionality lens to examine climate migration highlights the need for comprehensive and rights-based policies within collaborative climate action and networked climate governance. Policy implications include the protection of human rights, inclusion of climate migrants in decision-making processes, international cooperation and responsibility sharing, addressing root causes of migration, promoting sustainable livelihoods, and adopting gender-responsive approaches. By centering the needs and rights of climate migrants in policy frameworks, collaborative climate action can ensure the well-being and dignity of those affected by climate-induced displacement.

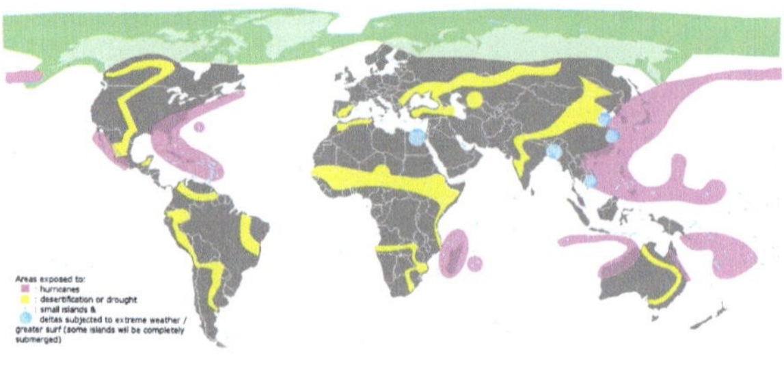

Figure 8.8. Major determinants of global climate migration.

Source: Wikipedia.

Climate migration in the face of a changing environment is an increasingly important issue (Figure 8.8 for the major factors). Climate change affects everyone, but it disproportionately impacts vulnerable and marginalized populations due to their lack of access to resources and assistance (Kumar & Singh, 2020). Intersecting social vulnerabilities such as gender, race, class or ethnicity can amplify climate-related displacement risk for these individuals (Bos et al., 2018; Kumar & Singh, 2020). Therefore, collective action must be taken by communities that are affected by climate risks along with governments at all levels to ensure that those most exposed—especially women and children—are protected from potential threats associated with environmental disruption (Höhne et al., 2019). Networked governance systems have been proposed as one way to effectively address this problem. Networked approaches focus on building relationships between government agencies through policy coordination within countries while also combining international efforts across borders (Palmiero et al., 2017). This type of system relies heavily upon collaboration among actors involved in addressing the challenges posed by climate change including civil society organizations, local authorities, public institutions and private companies who work together toward identifying solutions (Gonzáles-Vicente & Adell-Pérez, 2018).

Additionally collaborative initiatives like city networks or regional platforms provide a platform for people facing similar issues related to migration which could allow them greater opportunity to create unified strategies towards adaptation measures or community-based projects focused on resilience building before displacement becomes necessary (Stohl & Hagan Fernandez de Mera, 2018). By doing so networked climate governance has the potential both to protect those most vulnerable against additional threat posed by forced mobility while also allowing more flexibility in decision making processes when responding to localized changes resulting from global warming. This discussion highlights how complex decisions about managing population movement will need to take into account multiple factors if they are going to succeed over the longer term. One possible solution many suggest then lies in geoengineering – large scale interventions designed to mitigate effects of anthropogenic warming either directly cooling Earth's atmosphere artificially manipulating carbon cycle thus reducing emissions contributing greenhouse gasses causing rise temperatures globally).

8.5. LOOKING FORWARD TO THE POSSIBILITIES OF GEOENGINEERING

Using an intersectionality lens, it is crucial to critically discuss geoengineering and its policy implications within the context of collaborative climate action and networked climate governance (Bjerkan & Ryghaug, 2021; Fisher & Nasrin, 2021; Hitziger et al., 2019; Kelman & Clark-Ginsberg, 2022). Geoengineering refers to deliberate large-scale interventions in the Earth's systems to counteract climate change, such as solar radiation management and carbon dioxide removal. Examining geoengineering through an intersectional perspective allows us to understand its potential impacts on different social groups, taking into account factors such as race, gender, socioeconomic status, and nationality. One key aspect of addressing geoengineering in collaborative climate action and networked climate governance is the need for inclusive and participatory decision-making processes. Policies should ensure that diverse voices and perspectives, including those of marginalized communities, are included in the discussions and decisions related to geoengineering. It is crucial to recognize that different social groups may have different concerns, priorities, and levels of vulnerability to the potential risks and benefits of geoengineering technologies.

Moreover, an intersectional analysis sheds light on the potential uneven distribution of the impacts of geoengineering interventions (Hitziger et al., 2019; Hoffmann & Villamayor-Tomas, 2023; Homobono et al., 2022; Kellner, 2023). Certain regions and communities may be disproportionately affected by the implementation of geoengineering, while others may bear fewer risks or even benefit from these interventions. It is essential to consider the social, economic, and environmental justice implications of geoengineering policies and to prioritize the equitable distribution of both risks and benefits. Policy frameworks should also address the potential ethical concerns associated with geoengineering. An intersectional lens highlights the need to examine the power dynamics, accountability, and governance structures surrounding these technologies. Questions regarding who has the power to make decisions about geoengineering, who bears the responsibility for potential harms, and how decisions are made and monitored must be critically evaluated. Policies should aim to foster transparency, inclusivity, and accountability in the governance of geoengineering research and deployment. Figure 8.9 highlights some of the key aspects of geoengineering.

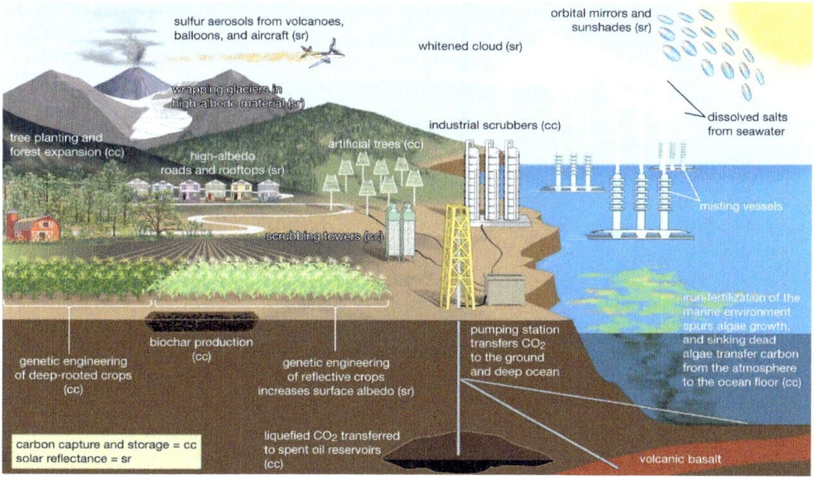

Figure 8.9. An overview of geoengineering.

Source: Britannica.

Additionally, considering the intersectionality of knowledge systems is crucial in the discussion of geoengineering. Indigenous knowledge and perspectives should be actively sought and respected in the evaluation and governance of geoengineering technologies (Rozance et al., 2019; Rudnick et al., 2019; Rutland & Aylett, 2008). Indigenous communities often possess valuable ecological knowledge and have a deep connection to the land, which can provide valuable insights into the potential impacts and risks associated with geoengineering. Inclusive governance should ensure that Indigenous rights, knowledge, and sovereignty are respected throughout the decision-making processes. Furthermore, an intersectional analysis emphasizes the importance of considering the interconnections between geoengineering and other social and environmental challenges. Geoengineering should not be viewed as a silver bullet solution to climate change but rather as part of a broader suite of climate mitigation and adaptation strategies. Policies should prioritize holistic approaches that address the root causes of climate change, such as reducing greenhouse gas emissions, promoting sustainable development, and fostering social and economic justice. Integration with other climate actions and the pursuit of sustainable development goals should be central in geoengineering policies. Lastly, an intersectional perspective highlights the need for ongoing research, monitoring, and assessment of geoengineering technologies. Policies should prioritize scientific research, international collaboration, and the establishment of robust monitoring and evaluation mechanisms to ensure that the potential risks, unintended consequences, and impacts of geoengineering are thoroughly assessed and understood.

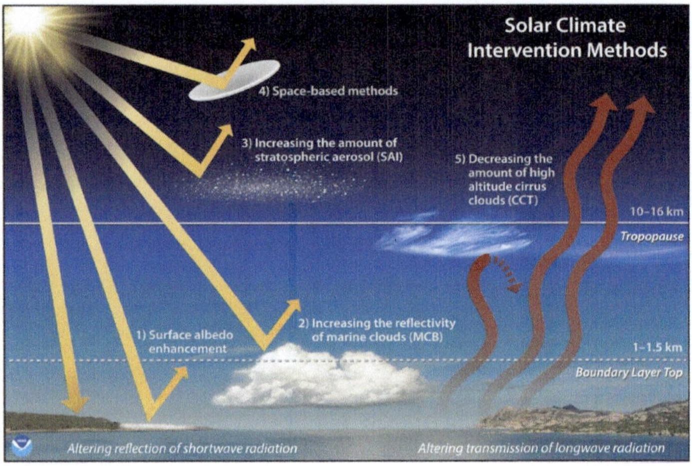

Figure 8.10. Illustration of solar geoengineering.

Source: Wikipedia.

A case in point is solar geoengineering (Figure 8.10). Solar geoengineering is a form of climate engineering that has been widely discussed as a possible solution to the global warming crisis. It involves deliberately reflecting some portion of incoming solar radiation back into space using various techniques, such as cloud seeding or deploying reflective aerosols in the upper atmosphere (Kravitz et al., 2019). Solar geoengineering can be an effective tool for mitigating climate change if done properly and with careful consideration. However, it also carries certain risks and uncertainties that need to be taken into account before implementation. The way in which solar geoengineering interacts with collective climate action and networked governance structures needs to consider intersectionality; this is because there are many different actors who may have differing goals, interests and perspectives when it comes to implementing these types of strategies (Garcia-Carreras et al., 2018). For instance, while nations like India may benefit from reduced temperatures due to deployed stratospheric aerosols under certain scenarios (Bala & Cao 2020), other countries could experience unwanted side effects such as changes in monsoon patterns or regional temperature increases. This demonstrates how important it is for decision makers to take into account multiple stakeholders' concerns when considering whether or not solar geoengineering should be used on a large scale. Furthermore, research suggests that women tend to be disproportionately affected by extreme weather events associated with anthropogenic climate change compared to men (Funk et al., 2018). Therefore, any policies related directly or indirectly targeting gender inequality must also factor in potential impacts from solar geoengineering initiatives so that all genders can benefit equally from them if they are implemented successfully. In addition, researchers have found disparities between rich countries and poorer ones regarding access to environmental information about projects like those involving solar engineering technologies (Bellamy et al., 2017). These discrepancies further highlight why special attention must be paid towards ensuring equitable participation among different groups within society during discussions around proposed solutions like Solar Geoengineering.

Generally, adopting an intersectionality lens to examine geoengineering emphasizes the importance of inclusive and participatory decision-making processes, equitable distribution of risks and benefits, consideration of ethical concerns, incorporation of Indigenous knowledge, integration with broader climate action, and ongoing research and assessment (Anderton & Setzer, 2018; Granderson, 2017; Hawkins et al., 2016; Kithiia, 2015; van Bers et al., 2019). By centering marginalized communities, addressing power imbalances, and fostering transparency and accountability, collaborative climate action and networked climate governance can navigate the complexities of geoengineering in a way that respects social and environmental justice principles. Geoengineering is the intentional large-scale manipulation of environmental processes with the aim of alleviating or preventing some aspects of climate change (Kravitz, 2019). It can be considered an example of collective action to address global warming, and increasingly it has been proposed as a viable solution to reduce greenhouse gasses in order to mitigate future temperature increases. As such, geoengineering intersects with multiple levels and disciplines within climate action and networked governance.

At its core, geoengineering entails interventions that alter Earth systems at various scales so as to counterbalance anthropogenic emissions (Dal Poz et al., 2022; Friedman et al., 2020; Harmon et al., 2023; Kelman & Clark-Ginsberg, 2022; Sultana et al., 2019). Types range from solar radiation management techniques like cloud seeding or space mirrors (Lenton et al., 2018) which intercept incoming sunlight before reaching the atmosphere; carbon dioxide removal methods including bioenergy combined with carbon capture and storage (BECCS) or direct air capture technologies; afforestation schemes designed for reforestation on previously deforested land; ocean fertilization experiments meant to stimulate phytoplankton growth (Adams

& Keith, 2020); upscaling agricultural productivity by using GMOs; among others. This variety speaks volumes about how complex this strategy really is – involving both natural sciences such as chemistry or biology but also social science elements related to policy analysis for implementation purposes. The effort required transcends countries' boundaries since any alterations made will affect not only local ecosystems but entire regions around them too.

Yet despite being a highly contested issue due mainly ethical concerns regarding potential risks posed by untested approaches, there is general consensus amongst experts today that research into these solutions should continue if we are serious about reducing long term effects of global warming worldwide (Kravitz et al., 2017). Consequently governments are starting programs dedicated exclusively towards developing different ways geoengineering could work best depending on their individual needs: For instance China invested over $500 million during 2016 alone into projects focused primarily on increasing forest coverage while India declared plans late last year intending to install floating artificial islands equipped with advanced solar panels off their coastline's coastlines in conjunction with wind turbines installed near each island – thus creating what they call "Renewable Energy Farms" (Chaudhuri & Roychowdhury 2021). Such initiatives exemplify how regional actors may join forces when tackling similar problems collectively rather than working independently through traditional state mechanisms like taxation policies which would most likely lead nowhere given current political constraints.

CHAPTER 8 SUMMARY

In conclusion, the final chapter on "Intersectionality in Climate Change Action and Networked Governance" highlights the critical importance of incorporating an intersectionality lens into climate change action and networked governance. Intersectionality provides a powerful framework for understanding and addressing the complex dynamics and interconnected systems of power, privilege, and oppression that shape climate change impacts and responses. By centering the experiences and voices of marginalized communities, recognizing the intersecting dimensions of identity, and addressing social, economic, and environmental justice, collaborative climate action and networked governance can become more effective, inclusive, and equitable. Throughout the chapter, we have explored the various dimensions of intersectionality and their implications for climate change action and governance. We have critically examined the realities of environmental racism, the role of women, Indigenous knowledge and natural resource management, climate migration, and geoengineering. In each of these areas, we have seen how intersectionality helps to uncover the underlying power dynamics, systemic inequalities, and unique vulnerabilities that shape climate change impacts and responses.

Key policy implications have emerged from our discussions. These include the recognition and protection of the rights of marginalized communities, the need for inclusive and participatory decision-making processes, the integration of diverse knowledge systems, the pursuit of social and economic justice, and the establishment of transparent and accountable governance structures. These policy implications provide a roadmap for building more just and effective climate change responses. Furthermore, the chapter has emphasized the importance of collaboration and networked governance in addressing climate change. The interconnectedness of social, political, and ecological systems requires collective action and cooperation among diverse stakeholders. Networked governance structures provide opportunities for knowledge exchange, coordination, and joint problem-solving. By fostering collaboration and building partnerships, networked governance enhances the capacity to address the multifaceted challenges of climate change and integrate diverse perspectives into decision-making processes.

The chapter's exploration of intersectionality has also underscored the significance of ongoing research, monitoring, and evaluation. As the understanding of intersectionality evolves, it is essential to continually assess and reassess the impacts of climate change on marginalized communities and the effectiveness of climate change policies and interventions. This iterative process allows for continuous learning, adaptation, and improvement in climate change action and networked governance. Looking ahead, it is clear that the principles of intersectionality should remain central to collective climate action and networked climate governance. Future efforts should continue to integrate intersectional perspectives into policy development, implementation, and evaluation. By understanding the complexities of power, privilege, and oppression, and by addressing the unique needs and experiences of marginalized communities, we can build more just, resilient, and sustainable responses to climate change.

CONCLUDING REMARKS

In the book "Collective Climate Action and Networked Climate Governance," the author delved deep into the pressing issue of climate change and examined the importance of collective action and networked governance in addressing this global challenge. Throughout the course of the book, several major themes have emerged, shedding light on the significance of these topics for undergraduate students. From the urgency of climate action to the role of youth engagement and the power of collective mobilization, these themes provide invaluable insights and guidance for young individuals seeking to make a meaningful impact on climate change. One of the most salient themes explored in the book is the urgency of climate action. The author convincingly argues that we are at a critical juncture in our fight against climate change, and immediate action is necessary to prevent catastrophic consequences. This urgency resonates strongly with undergraduate students, who are poised to inherit a world significantly impacted by climate change. Understanding the urgency of the situation empowers students to take action and encourages them to participate actively in initiatives that contribute to climate resilience.

Furthermore, the book emphasizes the role of youth engagement in climate governance. Young people play a crucial role in advocating for climate justice and pushing for transformative change. As undergraduate students, they possess a unique position within society, possessing the energy, passion, and capacity to drive positive change. By shedding light on the significance of youth engagement, the book inspires students to become active agents in climate governance, both within their educational institutions and broader communities. The author also highlights the transformative power of collective mobilization. This emphasizes the need for diverse actors, including individuals, communities, organizations, and governments, to come together and collaborate effectively to address climate change. Such a theme resonates strongly with undergraduate students, as they have the opportunity to form networks and communities during their college years. The book demonstrates that by fostering collaboration and leveraging collective resources, students can amplify their impact and bring about meaningful change. Another important theme explored in the book is the role of technology and innovation in climate governance. The author illustrates how advancements in technology have the potential to facilitate collective climate action. Undergraduate students, who are often at the forefront of technological advancements, can harness their skills and knowledge to develop innovative solutions that address climate challenges.

By highlighting the transformative potential of technology, the book encourages students to explore interdisciplinary approaches and leverage their expertise to create sustainable solutions.

Moreover, the book emphasizes the significance of education and awareness in fostering climate action. It elucidates the role of universities and educational institutions in equipping students with the knowledge and skills necessary to tackle climate change effectively. Undergraduate students have a unique opportunity to engage with interdisciplinary courses and research that can broaden their understanding of climate issues. The book underscores the relevance of such educational opportunities, encouraging students to actively seek out programs and initiatives that deepen their understanding of climate change and its complexities. Additionally, the author sheds light on the importance of equity and justice in climate governance. The book highlights the disproportionate impacts of climate change on vulnerable communities and stresses the need for inclusive approaches that prioritize justice and fairness. This theme resonates strongly with undergraduate students, who often have a strong sense of social justice and are committed to promoting equity. The book encourages students to consider the intersectionality of climate change and actively work towards inclusive solutions that address the needs of marginalized communities. Finally, the book emphasizes the significance of individual actions in collective climate action. While systemic changes are crucial, individual choices and behaviors also play a role in shaping sustainable societies. The author emphasizes the power of individual agency and provides practical suggestions for incorporating sustainable practices into daily lives. This theme is particularly relevant to undergraduate students, who are at a stage in life where they can make conscious choices that align with their values and contribute to a more sustainable future.

In conclusion, "Collective Climate Action and Networked Climate Governance" provides undergraduate students with invaluable insights into the pressing issue of climate change and the importance of collective action. The major themes explored in the book, including the urgency of climate action, youth engagement, collective mobilization, technology and innovation, education and awareness, equity and justice, and individual actions, resonate strongly with undergraduate students. By engaging with these themes, students are empowered to make a meaningful impact on climate change, both as individuals and as active participants in collective climate action. The book serves as a guiding compass, inspiring students to think critically, collaborate, and become advocates for a more sustainable and just future. It is a call to action for undergraduate students to embrace their agency, harness their knowledge and skills, and contribute to the global effort of combating climate change.

Bibliography

1. Alexander, S. M., Armitage, D., & Charles, A., (2015). Social networks and transitions to co-management in Jamaican marine reserves and small-scale fisheries [article]. *Global Environmental Change-Human and Policy Dimensions, 35*, 213–225. https://doi.org/10.1016/j.gloenvcha.2015.09.001.
2. Anderton, K., & Setzer, J., (2018). Subnational climate entrepreneurship: Innovative climate action in California and So Paulo [article]. *Regional Environmental Change, 18*(5), 1273–1284. https://doi.org/10.1007/s10113-017-1160-2.
3. Archer, D., (2016). Building urban climate resilience through community-driven approaches to development experiences from Asia [article]. *International Journal of Climate Change Strategies and Management, 8*(5), 654–669. https://doi.org/10.1108/ijccsm-03-2014-0035.
4. Atalan, H. N., (2019). Markets and collective action: A case study of traditional wheat varieties in Turkey [article]. *Journal of Economy Culture and Society,* (59), 13–30. https://doi.org/10.26650/jecs402676.
5. Bailey, I., MacGill, I., Passey, R., & Compston, H., (2012). The fall (and rise) of carbon pricing in Australia: A political strategy analysis of the carbon pollution reduction scheme [article]. *Environmental Politics, 21*(5), 691–711. https://doi.org/10.1080/09644016.2012.705066.
6. Bernstein, S., & Hoffmann, M., (2018). The politics of decarbonization and the catalytic impact of subnational climate experiments [article].

Policy Sciences, *51*(2), 189–211. https://doi.org/10.1007/s11077-018-9314-8.

7. Bixler, R. P., Epanchin-Niell, R. S., Brunson, M. W., Tarver, R. D., Sikes, B. A., McClure, M., & Aslan, C. E., (2023). How social and ecological characteristics shape transaction costs in polycentric wildfire governance: Insights from the sequoia-kings canyon ecosystem, California, USA [article]. *Ecology and Society*, *28*(1), 15, Article 34. https://doi.org/10.5751/es-13834-280134.
8. Bjerkan, K. Y., & Ryghaug, M., (2021). Diverging pathways to port sustainability: How social processes shape and direct transition work [article]. *Technological Forecasting and Social Change*, *166*, 11, Article 120595. https://doi.org/10.1016/j.techfore.2021.120595.
9. Blok, A., (2018). The politics of urban climate risks: Theoretical and empirical lessons from Ulrich Beck's methodological cosmopolitanism [article]. *Journal of Risk Research*, *21*(1), 41–55. https://doi.org/10.1080/13669877.2017.1359203.
10. Bodin, O., (2017). Collaborative environmental governance: Achieving collective action in social-ecological systems [review]. *Science*, *357*(6352), 659. https://doi.org/10.1126/science.aan1114.
11. Bodin, O., Nohrstedt, D., Baird, J., Summers, R., & Plummer, R., (2019). Working at the "speed of trust": Pre-existing and emerging social ties in wildfire responder networks in Sweden and Canada [article]. *Regional Environmental Change*, *19*(8), 2353–2364. https://doi.org/10.1007/s10113-019-01546-z.
12. Bowen, K. J., Cradock-Henry, N. A., Koch, F., Patterson, J., Hayha, T., Vogt, J., & Barbi, F., (2017). Implementing the "sustainable development goals": Towards addressing three key governance challenges-collective action, trade-offs, and accountability [review]. *Current Opinion in Environmental Sustainability*, *26, 27*, 90–96. https://doi.org/10.1016/j.cosust.2017.05.002.
13. Brandenberger, L., Ingold, K., Fischer, M., Schlapfer, I., & Leifeld, P., (2022). Boundary spanning through engagement of policy actors in multiple issues (sic)(sic)(sic)palabras clave [article]. *Policy Studies Journal*, *50*(1), 35–64. https://doi.org/10.1111/psj.12404.
14. Brewer, J. F., (2013). Making an environmental market, unmaking adaptive capacity: Species commodification in the New England groundfishery [article]. *Geoforum*, *50*, 172–181. https://doi.org/10.1016/j.geoforum.2013.08.006.

15. Brewer, J. F., Molton, K., Alden, R., & Guenther, C., (2017). Accountability, transformative learning, and alternate futures for New England groundfish catch shares [article]. *Marine Policy*, *80*, 113–122. https://doi.org/10.1016/j.marpol.2016.09.015.

16. Buchtmann, M., Wise, R., O'Connell, D., Crosweller, M., & Edwards, J., (2022). Reforming Australia's approach to hazards and disaster risk: National leadership, systems thinking, and inclusive conversations about vulnerability [article; early access]. *Disaster Prevention and Management*, 25. https://doi.org/10.1108/dpm-08-2022-0168.

17. Butler, J. R. A., Wise, R. M., Meharg, S., Peterson, N., Bohensky, E. L., Lipsett-Moore, G., Skewes, T. D., et al., (2022). 'Walking along with development': Climate resilient pathways for political resource curses [article]. *Environmental Science & Policy*, *128*, 228–241. https://doi.org/10.1016/j.envsci.2021.11.020.

18. Calliari, E., Michetti, M., Farnia, L., & Rarnieri, E., (2019). A network approach for moving from planning to implementation in climate change adaptation: Evidence from southern Mexico [article]. *Environmental Science & Policy*, *93*, 146–157. https://doi.org/10.1016/j.envsci.2018.11.025.

19. Chaudhury, A. S., Thornton, T. F., Helfgott, A., & Sova, C., (2017). Applying the robust adaptation planning (RAP) framework to Ghana's agricultural climate change adaptation regime [article]. *Sustainability Science*, *12*(5), 657–676. https://doi.org/10.1007/s11625-017-0462-0.

20. Chu, E. K., (2018). Urban climate adaptation and the reshaping of state-society relations: The politics of community knowledge and mobilization in Indore, India [article]. *Urban Studies*, *55*(8), 1766–1782. https://doi.org/10.1177/0042098016686509.

21. Cody, K. C., (2018). Flexible water allocations and rotational delivery combined adapt irrigation systems to drought [article]. *Ecology and Society*, *23*(2), 46, Article 47. https://doi.org/10.5751/es-10193-230247.

22. Currarini, S., Marchiori, C., & Tavoni, A., (2016). Network economics and the environment: Insights and perspectives [article]. *Environmental & Resource Economics*, *65*(1), 159–189. https://doi.org/10.1007/s10640-015-9953-6.

23. Daher, B., Hannibal, B., Portney, K. E., & Mohtar, R. H., (2019). Toward creating an environment of cooperation between water, energy, and food stakeholders in San Antonio [article]. *Science of*

the Total Environment, *651*, 2913–2926. https://doi.org/10.1016/j.scitotenv.2018.09.395.

24. Dal Poz, M. E. S., Ignacio, P. S. D., Azevedo, A., Francisco, E. C., Piolli, A. L., da Silva, G. G., & Ribeiro, T. P., (2022). Food, energy and water nexus: An urban living laboratory development for sustainable systems transition [article]. *Sustainability*, *14*(12), 22, Article 7163. https://doi.org/10.3390/su14127163.
25. Davies, K., Fisher, K., Foley, M., Greenaway, A., Hewitt, J., Le Heron, R., Mikaere, H., et al., (2018). Navigating collaborative networks and cumulative effects for Sustainable Seas [article]. *Environmental Science & Policy*, *83*, 22–32. https://doi.org/10.1016/j.envsci.2018.01.013.
26. De Macedo, L. S. V., & Jacobi, P. R., (2019). Subnational politics of the urban age: Evidence from Brazil on integrating global climate goals in the municipal agenda [article]. *Palgrave Communications*, *5*, 15, Article 18. https://doi.org/10.1057/s41599-019-0225-x.
27. Deka, N., & Goswami, K., (2022). Organic cultivation and sustainable value chain development for tea smallholders: Findings from Assam, India [article]. *Sustainable Production and Consumption*, *32*, 562–579. https://doi.org/10.1016/j.spc.2022.05.004.
28. Den, U. R. M., & Russel, D. J., (2018). Climate adaptation in fragmented governance settings: The consequences of reform in public administration [article]. *Environmental Politics*, *27*(2), 341–361. https://doi.org/10.1080/09644016.2017.1386341.
29. Dey, A., Gupta, A., & Singh, G., (2017). Open innovation at different levels for higher climate risk resilience [article]. *Science Technology and Society*, *22*(3), 388–406. https://doi.org/10.1177/0971721817723242.
30. Diop, M., & Remvikos, Y., (2016). Dakar suburb communities and recurrent flooding: The relevance of social capital and adaptive capacity [article]. *Environnement Risques & Sante*, *15*(4), 341–350. https://doi.org/10.1684/ers.2016.0891.
31. Ensor, J., & Harvey, B., (2015). Social learning and climate change adaptation: Evidence for international development practice [article]. *Wiley Interdisciplinary Reviews-Climate Change*, *6*(5), 509–522. https://doi.org/10.1002/wcc.348.
32. Fidelis, T., Teles, F., Roebeling, P., & Riazi, F., (2019). Governance for sustainability of estuarine areas assessing alternative models using the case of Ria de Aveiro, Portugal [article]. *Water*, *11*(4), 18, Article 846.

https://doi.org/10.3390/w11040846.

33. Fisher, D. R., & Nasrin, S., (2021). Shifting coalitions within the youth climate movement in the US [article]. *Politics and Governance*, *9*(2), 112–123. https://doi.org/10.17645/pag.v9i2.3801.
34. Fitch-Roy, O., Fairbrass, J., & Benson, D., (2020). Ideas, coalitions and compromise: Reinterpreting EU-ETS lobbying through discursive institutionalism [article]. *Journal of European Public Policy*, *27*(1), 82–101. https://doi.org/10.1080/13501763.2019.1567573.
35. Fried, H. S., Hamilton, M., & Berardo, R., (2022). Closing integrative gaps in complex environmental governance systems [article]. *Ecology and Society*, *27*(1), 22, Article 15. https://doi.org/10.5751/es-12996-270115.
36. Friedman, R. S., Guerrero, A. M., McAllister, R. R. J., Rhodes, J. R., Santika, T., Budiharta, S., Indrawan, T., et al., (2020). Beyond the community in participatory forest management: A governance network perspective [article]. *Land Use Policy*, *97*, 11, Article 104738. https://doi.org/10.1016/j.landusepol.2020.104738.
37. Goodwin, D., Holman, I., Sutcliffe, C., Salmoral, G., Pardthaisong, L., Visessri, S., Ekkawatpanit, C., & Rey, D., (2022). The contribution of a catchment-scale advice network to successful agricultural drought adaptation in Northern Thailand [article]. *Philosophical Transactions of the Royal Society a-Mathematical Physical and Engineering Sciences*, *380*(2238), 21, Article 20210293. https://doi.org/10.1098/rsta.2021.0293.
38. Gordon, D. J., (2016). Lament for a network? Cities and networked climate governance in Canada [article]. *Environment and Planning C-Government and Policy*, *34*(3), 529–545. https://doi.org/10.1177/0263774x15614675.
39. Gore, C. D., (2010). The limits and opportunities of networks: Municipalities and Canadian climate change policy [article]. *Review of Policy Research*, *27*(1), 27–46. https://doi.org/10.1111/j.1541-1338.2009.00425.x.
40. Granderson, A. A., (2017). The role of traditional knowledge in building adaptive capacity for climate change: Perspectives from Vanuatu [article]. *Weather Climate and Society*, *9*(3), 545–561. https://doi.org/10.1175/wcas-d-16-0094.1.
41. Gregg, J. S., Bolwig, S., Sciullo, A., Arrobbio, O., Hubert, W., Ivask,

N., Iturriza, I. J., et al., (2023). How can energy become a community endeavor in Europe? Consortium benchmarking strategies for the mobilization of collective action initiatives [article]. *Energy Research & Social Science*, *98*, 12, Article 103005. https://doi.org/10.1016/j.erss.2023.103005.

42. Gronow, A., Wagner, P., & Yla-Anttila, T., (2020). Explaining collaboration in consensual and conflictual governance networks [article]. *Public Administration*, *98*(3), 730–745. https://doi.org/10.1111/padm.12641.
43. Grundig, F., & Ward, H., (2015). Structural group leadership and regime effectiveness [article]. *Political Studies*, *63*(1), 221–239. https://doi.org/10.1111/1467-9248.12056.
44. Gunningham, N., (2012). Confronting the challenge of energy governance [article]. *Transnational Environmental Law*, *1*(1), 119–135. https://doi.org/10.1017/s2047102511000124.
45. Hale, T., & Roger, C., (2014). Orchestration and transnational climate governance [article]. *Review of International Organizations*, *9*(1), 59–82. https://doi.org/10.1007/s11558-013-9174-0.
46. Hamilton, M., & Lubell, M., (2018). Collaborative governance of climate change adaptation across spatial and institutional scales [article]. *Policy Studies Journal*, *46*(2), 222–247. https://doi.org/10.1111/psj.12224.
47. Harmon, G., Jepson, W., & Lefore, N., (2023). Farmer-led irrigation development in Sub-Saharan Africa [article]. *Wiley Interdisciplinary Reviews-Water*, *10*(2), 13. https://doi.org/10.1002/wat2.1631.
48. Hawkins, C. V., Krause, R. M., Feiock, R. C., & Curley, C., (2016). Making meaningful commitments: Accounting for variation in cities' investments of staff and fiscal resources to sustainability [article]. *Urban Studies*, *53*(9), 1902–1924. https://doi.org/10.1177/0042098015580898.
49. Hawkins, C. V., Krause, R., Feiock, R. C., & Curley, C., (2018). The administration and management of environmental sustainability initiatives: A collaborative perspective [article]. *Journal of Environmental Planning and Management*, *61*(11), 2015–2031. https://doi.org/10.1080/09640568.2017.1379959.
50. Heikkila, T., & Gerlak, A. K., (2019). Working on learning: How the institutional rules of environmental governance matter [article]. *Journal of Environmental Planning and Management*, *62*(1), 106–123. https://doi.org/10.1080/09640568.2018.1473244.

51. Henry, A. D., & Vollan, B., (2014). Networks and the challenge of sustainable development. In: Gadgil, A., & Liverman, D. M., (eds.), *Annual Review of Environment and Resources* (Vol. 39, pp. 583–610). Annual Reviews. https://doi.org/10.1146/annurev-environ-101813-013246.

52. Herzog, L. M., & Ingold, K., (2019). Threats to common-pool resources and the importance of forums: On the emergence of cooperation in CPR problem settings [article]. *Policy Studies Journal*, *47*(1), 77–113. https://doi.org/10.1111/psj.12308.

53. Hitziger, M., Aragrande, M., Berezowski, J. A., Canali, M., Vilas, V. D., Hoffmann, S., Igrejas, G., et al., (2019). Evolvinc: Evaluating knowledge integration capacity in multistakeholder governance [article]. *Ecology and Society*, *24*(2), 16, Article 36. https://doi.org/10.5751/es-10935-240236.

54. Hoffmann, P., & Villamayor-Tomas, S., (2023). Irrigation modernization and the efficiency paradox: A meta-study through the lens of networks of action situations [article]. *Sustainability Science*, *18*(1), 181–199. https://doi.org/10.1007/s11625-022-01136-9.

55. Homobono, T., Guimaraes, M. H., Esgalhado, C., & Madureira, L., (2022). Water governance in Mediterranean farming systems through the social-ecological systems framework-an empirical case in Southern Portugal [article]. *Land*, *11*(2), 18, Article 178. https://doi.org/10.3390/land11020178.

56. Kellner, E., (2023). Identifying leverage points for shifting water-energy-food nexus cases towards sustainability through the networks of action situations approach combined with systems thinking [article]. *Sustainability Science*, *18*(1), 135–152. https://doi.org/10.1007/s11625-022-01170-7.

57. Kelman, I., & Clark-Ginsberg, A., (2022). An urban governance framework for including environmental migrants in sustainable cities [article]. *Climate*, *10*(8), 16, Article 121. https://doi.org/10.3390/cli10080121.

58. Kim, Y. S., Bae, J. S., Fisher, L. A., & Jung, J., (2019). Branding K-REDD plus and its implications for South Korea's middle power diplomacy [article]. *Pacific Review*, *32*(1), 76–89. https://doi.org/10.1080/09512748.2018.1428674.

59. Kithiia, J., (2015). Resourceless victims or resourceful collectives: Addressing the impacts of climate change through social capital in

fringing coastal communities [article]. *Ocean & Coastal Management, 106*, 110–117. https://doi.org/10.1016/j.ocecoaman.2015.01.020.

60. Koefoed, O., (2019). Urban nature as transformed practice – A case of multi-dimensional processing to increase public value in Copenhagen [article]. *Local Economy, 34*(6), 525–544. https://doi.org/10.1177/0269094219882670.
61. Koontz, T. M., Gupta, D., Mudliar, P., & Ranjan, P., (2015). Adaptive institutions in social-ecological systems governance: A synthesis framework [article]. *Environmental Science & Policy, 53*, 139–151. https://doi.org/10.1016/j.envsci.2015.01.003.
62. Krause, R. M., Hawkins, C. V., & Park, A. Y. S., (2021). The perfect amount of help: An examination of the relationship between capacity and collaboration in urban energy and climate initiatives [article]. *Urban Affairs Review, 57*(2), 583–608, Article 1078087419884650. https://doi.org/10.1177/1078087419884650.
63. Laycock, K. E., & Mitchell, C. L., (2019). Social capital and incremental transformative change: Responding to climate change experts in Metro Manila [article]. *Climatic Change, 152*(1), 47–66. https://doi.org/10.1007/s10584-018-2360-6.
64. Lee, E., Jung, C. S., & Lee, M. K., (2014). The potential role of boundary organizations in the climate regime [article]. *Environmental Science & Policy, 36*, 24–36. https://doi.org/10.1016/j.envsci.2013.07.008.
65. Lee, T., & Koski, C., (2014). Mitigating global warming in global cities: Comparing participation and climate change policies of C40 cities [article]. *Journal of Comparative Policy Analysis, 16*(5), 475–492. https://doi.org/10.1080/13876988.2014.910938.
66. Leventon, J., Schaal, T., Velten, S., Danhardt, J., Fischer, J., Abson, D. J., & Newig, J., (2017). Collaboration or fragmentation? Biodiversity management through the common agricultural policy [article]. *Land Use Policy, 64*, 1–12. https://doi.org/10.1016/j.landusepol.2017.02.009.
67. Lubell, M., & Robbins, M., (2022). Adapting to sea-level rise: Centralization or decentralization in polycentric governance systems?(sic)(sic)(sic)palabras clave [article]. *Policy Studies Journal, 50*(1), 143–175. https://doi.org/10.1111/psj.12430.
68. Luthe, T., & Wyss, R., (2016). Resilience to climate change in a cross-scale tourism governance context: A combined quantitative-qualitative network analysis [article]. *Ecology and Society, 21*(1), 18, Article 27.

https://doi.org/10.5751/es-08234-210127.

69. Mar, K. A., Schapke, N., Fraude, C., Bruhn, T., Wamsler, C., Stasiak, D., Schroeder, H., & Lawrence, M. G., (2023). Learning and community building in support of collective action: Toward a new climate of communication at the COP [article; early access]. *Wiley Interdisciplinary Reviews-Climate Change*, 12. https://doi.org/10.1002/wcc.832.
70. McLaughlin, D. M., (2022). Evaluating the impact of different spatial linkages on forum outcome interdependencies in polycentric systems [article; early access]. *Public Administration Review*, 18. https://doi.org/10.1111/puar.13599.
71. Medina-Garcia, C., Nagarajan, S., Castillo-Vysokolan, L., Beatse, E., & Van, D. B. P., (2022). Innovative Multi-actor collaborations as collective actors and institutionalized spaces. The case of food governance transformation in Leuven (Belgium) [article]. *Frontiers in Sustainable Food Systems*, *5*, 23, Article 788934. https://doi.org/10.3389/fsufs.2021.788934.
72. Morrison, T. H., Adger, W. N., Brown, K., Lemos, M. C., Huitema, D., & Hughes, T. P., (2017). Mitigation and adaptation in polycentric systems: Sources of power in the pursuit of collective goals [review]. *Wiley Interdisciplinary Reviews-Climate Change*, *8*(5), 16, Article e479. https://doi.org/10.1002/wcc.479.
73. National Intelligence Council, (2017). The US national intelligence council on global demographic trends [editorial material]. *Population and Development Review*, *43*(2), 387–391. https://doi.org/10.1111/padr.12071.
74. Neville, K. J., Cook, J., Baka, J., Bakker, K., & Weinthal, E. S., (2019). Can shareholder advocacy shape energy governance? The case of the US antifracking movement [article]. *Review of International Political Economy*, *26*(1), 104–133. https://doi.org/10.1080/09692290.2018.1488757.
75. Niles, M. T., & Lubell, M., (2012). Integrative frontiers in environmental policy theory and research [review]. *Policy Studies Journal*, *40*, 41–64. https://doi.org/10.1111/j.1541-0072.2012.00445.x.
76. Niles, M. T., Rudnick, J., Lubell, M., & Cramer, L., (2021). Household and community social capital links to smallholder food security [article]. *Frontiers in Sustainable Food Systems*, *5*, 14, Article 583353. https://doi.org/10.3389/fsufs.2021.583353.

77. O'Brien, L., Karlsdottir, B., Ambrose-Oji, B., Urquhart, J., Edwards, D., Amboage, R., & Jones, G., (2021). How can local and regional knowledge networks contribute to landscape level action for tree health? [article]. *Forests*, *12*(10), 17, Article 1394. https://doi.org/10.3390/f12101394.

78. Olivier, T., & Berardo, R., (2022). Birds of a feather fight together: Forum involvement in a weakly institutionalized ecology of policy games(sic)(sic)(sic)Palabras clave [article]. *Policy Studies Journal*, *50*(1), 176–198. https://doi.org/10.1111/psj.12418.

79. Orach, K., & Schluter, M., (2016). Uncovering the political dimension of social-ecological systems: Contributions from policy process frameworks [article]. *Global Environmental Change-Human and Policy Dimensions*, *40*, 13–25. https://doi.org/10.1016/j.gloenvcha.2016.06.002.

80. Ostrom, E., (2014). A polycentric approach for coping with climate change [article]. *Annals of Economics and Finance*, *15*(1), 97–134. <Go to ISI>://WOS:000337200300003.

81. Panditharatne, C., (2016). Institutional barriers in adapting to climate change: A case study in Sri Lanka [article]. *Ocean & Coastal Management*, *130*, 73–78. https://doi.org/10.1016/j.ocecoaman.2016.06.003.

82. Partelow, S., Schluter, A., Armitage, D., Bavinck, M., Carlisle, K., Gruby, R. L., Hornidge, A. K., et al., (2020). Environmental governance theories: A review and application to coastal systems [review]. *Ecology and Society*, *25*(4), 21, Article 19. https://doi.org/10.5751/es-12067-250419.

83. Pavanelli, J. M. M., Sang, E. V., De Oliveira, C. E., Campos, F. D., Lazaro, L. L. B., Edomah, N., & Igari, A. T., (2023). An institutional framework for energy transitions: Lessons from the Nigerian electricity industry history [article]. *Energy Research & Social Science*, *97*, 15, Article 102994. https://doi.org/10.1016/j.erss.2023.102994.

84. Perz, S. G., Arteaga, M., Farah, A. B., Brown, I. F., Mendoza, E. R. H., De Paula, Y. A. P., Yabar, L. M. P., et al., (2022). Participatory action research for conservation and development: Experiences from the amazon [article]. *Sustainability*, *14*(1), 29, Article 233. https://doi.org/10.3390/su14010233.

85. Plummer, R., Baird, J., Farhad, S., & Witkowski, S., (2020). How do biosphere stewards actively shape trajectories of social-ecological

change? [article]. *Journal of Environmental Management*, *261*, 6, Article 110139. https://doi.org/10.1016/j.jenvman.2020.110139.

86. Rahimi-Feyzabad, F., Yazdanpanah, M., Gholamrezai, S., & Ahmadvand, M., (2022). Social network analysis of institutions involved in groundwater resources management: Lessons learned from Iran [article]. *Journal of Hydrology*, *613*, 15, Article 128442. https://doi.org/10.1016/j.jhydrol.2022.128442.
87. Rashchupkina, Y., (2015). The European Union's role in networks on removal of fossil fuel subsidies and disclosure of climate change information [article]. *Contemporary Politics*, *21*(3), 354–366. https://doi.org/10.1080/13569775.2015.1061244.
88. Rozance, M. A., Denton, A., Matsler, A. M., Grabowski, Z., & Mayhugh, W., (2019). Examining the scalar knowledge politics of risk within coastal sea level rise adaptation planning knowledge systems [article]. *Environmental Science & Policy*, *99*, 105–114. https://doi.org/10.1016/j.envsci.2019.05.024.
89. Rudnick, J., Niles, M., Lubell, M., & Cramer, L., (2019). A comparative analysis of governance and leadership in agricultural development policy networks [article]. *World Development*, *117*, 112–126. https://doi.org/10.1016/j.worlddev.2018.12.015.
90. Rutland, T., & Aylett, A., (2008). The work of policy: Actor networks, governmentality, and local action on climate change in Portland, Oregon [article]. *Environment and Planning D-Society & Space*, *26*(4), 627–646. https://doi.org/10.1068/d6907.
91. Saint, V. A. S., Hickey, G. M., Locher, U., & Phillip, L. E., (2016). Exploring the role of social capital in influencing knowledge flows and innovation in smallholder farming communities in the Caribbean [article]. *Food Security*, *8*(3), 535–549. https://doi.org/10.1007/s12571-016-0581-y.
92. Salmoral, G., Zegarra, E., Vazquez-Rowe, I., Gonzalez, F., Del Castillo, L., Saravia, G. R., Graves, A., et al., (2020). Water-related challenges in nexus governance for sustainable development: Insights from the city of Arequipa, Peru [article]. *Science of the Total Environment*, *747*, 13, Article 141114. https://doi.org/10.1016/j.scitotenv.2020.141114.
93. Sanders, A. J. P., Hyldmo, H. D., Prasti, H. R. D., Ford, R. M., Larson, A. M., & Keenan, R. J., (2017). Guinea pig or pioneer: Translating global environmental objectives through to local actions in Central Kalimantan, Indonesia's REDD plus pilot province [article]. *Global*

Environmental Change-Human and Policy Dimensions, *42*, 68–81. https://doi.org/10.1016/j.gloenvcha.2016.12.003.

94. Sara, L. M., Jameson, S., Pfeffer, K., & Baud, I., (2016). Risk perception: The social construction of spatial knowledge around climate change-related scenarios in lima [article]. *Habitat International*, *54*, 136–149. https://doi.org/10.1016/j.habitatint.2015.12.025.
95. Sattler, C., & Schroter, B., (2022). Collective action across boundaries: Collaborative network initiatives as boundary organizations to improve ecosystem services governance [article]. *Ecosystem Services*, *56*, 20, Article 101452. https://doi.org/10.1016/j.ecoser.2022.101452.
96. Sayegh, A. G., (2020). Moral duties, compliance and polycentric climate governance [article]. *International Environmental Agreements-Politics Law and Economics*, *20*(3), 483–506. https://doi.org/10.1007/s10784-020-09494-4.
97. Shrestha, G., Pakhtigian, E. L., & Jeuland, M., (2023). Women who do not migrate: Intersectionality, social relations, and participation in Western Nepal [article]. *World Development*, *161*, 17, Article 106109. https://doi.org/10.1016/j.worlddev.2022.106109.
98. Sultana, P., Thompson, P. M., Paudel, N. S., Pariyar, M., & Rahman, M., (2019). Transforming local natural resource conflicts to cooperation in a changing climate: Bangladesh and Nepal lessons [article]. *Climate Policy*, *19*, S94-S106. https://doi.org/10.1080/14693062.2018.1527678.
99. Teodoro, J. D., Prell, C., & Sun, L. X., (2021). Quantifying stakeholder learning in climate change adaptation across multiple relational and participatory networks [article]. *Journal of Environmental Management*, *278*, 10, Article 111508. https://doi.org/10.1016/j.jenvman.2020.111508.
100. Therville, C., Brady, U., Barreteau, O., Bousquet, F., Mathevet, R., Dhenain, S., Grelot, F., & Bremond, P., (2019). Challenges for local adaptation when governance scales overlap. Evidence from Languedoc, France [article]. *Regional Environmental Change*, *19*(7), 1865–1877. https://doi.org/10.1007/s10113-018-1427-2.
101. Tosun, J., & Schoenefeld, J. J., (2017). Collective climate action and networked climate governance [article]. *Wiley Interdisciplinary Reviews-Climate Change*, *8*(1), 17, Article e440. https://doi.org/10.1002/wcc.440.
102. Tuda, A. O., & Machumu, M. E., (2019). Institutions and adaptive

capacity for marine biodiversity conservation [article]. *Environmental Science & Policy, 100,* 238–246. https://doi.org/10.1016/j.envsci.2019.03.012.

103. Van Popering-Verkerk, J., Molenveld, A., Duijn, M., Van, L. C., & Van, B. A., (2022). A framework for governance capacity: A broad perspective on steering efforts in society [article]. *Administration & Society, 54*(9), 1767–1794, Article 00953997211069932. https://doi.org/10.1177/00953997211069932.
104. Van, B. A., & Eshuis, J., (2009). *Knowledge Governance: Complementing Hierarchies, Networks and Markets? [knowledge Democracy: Consequences for Science, Politics, and Media].* International Conference Towards Knowledge Democracy, Leiden, Netherlands.
105. Van, B. C., Delaney, A., Eakin, H., Cramer, L., Purdon, M., Oberlack, C., Evans, T., et al., (2019). Advancing the research agenda on food systems governance and transformation [review]. *Current Opinion in Environmental Sustainability, 39,* 94–102. https://doi.org/10.1016/j.cosust.2019.08.003.
106. Vantaggiato, F., Lubell, M., Hummel, M., Chow, A. C. H., & Siwe, A. T., (2023). Creating adaptive social-ecological fit: The role of regional actors in the governance of Sea-level rise adaptation in San Francisco Bay [article]. *Global Environmental Change-Human and Policy Dimensions, 80,* 12, Article 102654. https://doi.org/10.1016/j.gloenvcha.2023.102654.
107. Vasconcelos, V. V., Santos, F. C., & Pacheco, J. M., (2015). Cooperation dynamics of polycentric climate governance [article]. *Mathematical Models & Methods in Applied Sciences, 25*(13), 2503–2517. https://doi.org/10.1142/s0218202515400163.
108. Virah-Sawmy, M., Duran, A. P., Green, J. M. H., Guerrero, A. M., Biggs, D., & West, C. D., (2019). Sustainability gridlock in a global agricultural commodity chain: Reframing the soy-meat food system [article]. *Sustainable Production and Consumption, 18,* 210–223. https://doi.org/10.1016/j.spc.2019.01.003.
109. Voogd, R., Rudberg, P. M., De Vries, J. R., Beunen, R., Espiritu, A. A., Methner, N., Larsen, R. K., et al., (2022). A systematic review on the role of trust in the water governance literature [review]. *Water Research X, 16,* 12, Article 100147. https://doi.org/10.1016/j.wroa.2022.100147.
110. Watabe, A., & Gilby, S., (2020). To see a world in a grain of sand-

the transformative potential of small community actions [article]. *Sustainability, 12*(18), 19, Article 7404. https://doi.org/10.3390/su12187404.

111. Wernli, D., Bottcher, L., Vanackere, F., Kaspiarovich, Y., Masood, M., & Levrat, N., (2023). Understanding and governing global systemic crises in the 21st century: A complexity perspective [article; early access]. *Global Policy*, 22. https://doi.org/10.1111/1758-5899.13192.
112. Willems, J. J., & Giezen, M., (2022). Understanding the institutional work of boundary objects in climate-proofing cities: The case of Amsterdam rainproof [article]. *Urban Climate*, *44*, 13, Article 101222. https://doi.org/10.1016/j.uclim.2022.101222.
113. Wyborn, C., (2015). Cross-scale linkages in connectivity conservation: Adaptive governance challenges in spatially distributed networks [article]. *Environmental Policy and Governance*, *25*(1), 1–15. https://doi.org/10.1002/eet.1657.
114. Yates, J. S., (2012). Uneven interventions and the scalar politics of governing livelihood adaptation in rural Nepal [article]. *Global Environmental Change-Human and Policy Dimensions*, *22*(2), 537–546. https://doi.org/10.1016/j.gloenvcha.2012.01.007.
115. Yoder, L., Wagner, C. H., Sullivan-Wiley, K., & Smith, G., (2022). The promise of collective action for large-scale commons dilemmas: Reflections on common-pool-resource theory [review]. *International Journal of the Commons*, *16*(1), 47–63. https://doi.org/10.5334/ijc.1163.
116. Zeren, G. B., Tan, J., Zhang, Q., & Qiuying, B., (2022). Rebuilding rural community cooperative institutions and their role in herder adaptation to climate change [article; early access]. *Climate Policy*, 16. https://doi.org/10.1080/14693062.2023.2184763.
117. Zhan, M. X., (2022). Sustainable consumption and the well-being dividend: Insights from the zero-waste movement in Chinese cities [article]. *Sustainability-Science Practice and Policy*, *18*(1), 731–748. https://doi.org/10.1080/15487733.2022.2111925.

INDEX

F

H

I

J

K

L

G

T

U

V